SAFETY ANALYSIS

SAFETY ANALYSIS
PRINCIPLES AND PRACTICE IN OCCUPATIONAL SAFETY

Lars Harms-Ringdahl

CRC Press
Taylor & Francis Group
Boca Raton London New York

CRC Press is an imprint of the
Taylor & Francis Group, an **informa** business

CRC Press
Taylor & Francis Group
6000 Broken Sound Parkway NW, Suite 300
Boca Raton, FL 33487-2742

First issued in paperback 2019

© 2001 by Taylor & Francis Group, LLC
CRC Press is an imprint of Taylor & Francis Group, an Informa business

No claim to original U.S. Government works

ISBN-13: 978-0-415-23655-3 (hbk)
ISBN-13: 978-0-367-39718-0 (pbk)

Library of Congress Cataloging-in-Publication Data

Catalog record is available from the Library of Congress

Visit the Taylor & Francis Web site at
http://www.taylorandfrancis.com

and the CRC Press Web site at
http://www.crcpress.com

Contents

Preface

Over a number of years, great interest has been shown in the prevention of accidents that may have major consequences. This applies above all to technologically advanced installations in the chemicals processing and nuclear industries. A great deal of effort has been put in, and much research and practical work on development has been devoted to how major accidents can be prevented. Safety analysis has become a methodology that is applied to a growing extent, often providing the basis for safety activities at plant level.

Occupational accidents are another serious problem, even greater than major hazards. In the world as a whole, the International Labour Organization (ILO) estimates that around 300 000 people are killed and 250 million injured in occupational accidents each year (Takala, 1998). Given the scale of the problem, this area actually deserves greater attention than that paid to major accidents. It is therefore essential to take advantage of the assistance that safety analysis can provide in preventing common accidents at work.

The aim of this book is to describe practical approaches and methods for safety analysis, especially for applications in the regular occupational environment. The idea is to give simple straightforward descriptions and relate practical experiences to promote wider use of the methodology.

Safety analysis as a tool

The basic perspective of this book is that safety analysis is a tool that can be employed in safety work. By utilising appropriate methods, the knowledge that is available in a workplace can be supplemented and applied more systematically.

Safety analysis is a systematic procedure for analysing systems to identify and evaluate hazards and safety characteristics. A safety analysis usually has three main elements: identification of hazards, assessment of the risks that arise, and the generation of measures that can increase the level of safety.

Safety analysis is one tool among others. It is not a shortcut but represents part of safety work as a whole. It is not particularly difficult, nor is it remarkable or peculiar in any sense. If an analysis is suitably designed, good results can be obtained from a few hours or a few days work.

Safety analysis is currently employed in the arena of occupational accident prevention to only a small extent. But there are several methods available, and experiences from applying them are favourable. Risks can be reduced, and systematic safety work has economic benefits—for both companies and society.

A further argument is that the increased complexity of modern production systems means that a systematic approach to safety activities is indispensable. Otherwise, risks cannot be handled rationally and efficiently.

Organisation of the book

The idea of the book is to show how safety analysis can be practically applied in the field of occupational safety. Its emphasis lies on explaining how different methods work.

A large number of different methods—around 50—are referred to. Three selected methods with a focus on occupational accidents are described in detail: Energy Analysis, Job Safety Analysis and Deviation Analysis. To introduce the reader to a broader range of possible methods, a number of other approaches are also described. These include Hazard and Operability Studies (HAZOP) and Fault Tree Analysis. A rather extensive overview of the methods is given in Chapter 12. The main focus is on qualitative methods, since it is the author's experience that probabilistic methods have less applicability in this context.

Another major theme concerns the analytical procedure, i.e. the various stages that make up an analysis and how these are related to one another. This procedure includes the planning of an analysis. Some views are also expressed on how high quality analyses can be achieved.

The aim of an analysis is to obtain results—in the form of reduced risk for occupational accidents. An analysis should be seen as a supplement to a company's own safety activities. Arguments for and against conducting a safety analysis are presented. One section is devoted to economic appraisals of safety analyses. Five examples are provided, and the results show that safety analysis tends to be a profitable as well as socially desirable activity.

The idea is that the book will function as a guide for people who would like to employ the methodology. For this reason, there is no stress on theory. A bibliography is provided for those who want to go further.

The book was originally written with safety practitioners and the Labour Inspectorate in mind, but it should also be of interest to anyone involved in occupational injury prevention. Designers of machines and workplaces are some of the persons in mind. The book also provides an orientation in safety analysis to potential commissioners of an analysis from a consultant, or safety committee members who want to be able to question management on what methods have been used to assess risks to comply with legislation. For such persons the methodological overview and advice on planning might be most relevant (chapters 12 and 13).

The material is derived from both the specialised literature and the author's own work on the analysis and prevention of accidents. Certain important lessons have been learned from training courses in safety analysis. The experiences of course participants have led to improved accounts of methods and the various ways in which use is made of results. Experiences from these courses suggest that safety analysis is a procedure that should be more widely adopted, e.g. by safety engineers.

This second English edition contains an enlarged account of the methods involved. Parts of the first edition not directly concerned with methods have been removed. This new edition represents a translated, revised and extended version of a book that first came out in Swedish (Harms-Ringdahl, 1987b).

The word "company" comes up throughout the book, and this should be interpreted in a wide sense. It refers to any type of organised production where there is any type of physical risk. It does not imply a limitation to the private sector.

The book is primarily concerned with occupational accidents, but the methods and approach can also be applied in many other areas and to different types of systems. There is some discussion of wider forms of application, referred to here as integrated approaches to hazard identification and risk assessment.

Persons most interested in practical applications can skip over several of the chapters. It might be appropriate for them to start with chapters 3 to 7, and then quickly review the contents of Chapter 13. Often it is only when you have tried yourself that the benefits of safety analysis become apparent.

Acknowledgements

The material in this book has been developed over several years through co-operation and contacts with many people. My thanks go to research colleagues and friends in Sweden and other countries, and to the many others who have encouraged the application and development of safety analysis in an occupational context.

I gratefully acknowledge the permission to reproduce material on HAZOP supplied by the United Kingdom Chemical Industries Association Ltd. Finally, I thank the Swedish Council for Work Life Research for the financial support they have provided for many of the projects on which this book is based.

1
Accidents and safety

1.1 THE ACCIDENT PROBLEM

Introduction

Risks and accidents are serious problems from many different perspectives. Some of these are discussed as background to the need for accident prevention and related tools. This section takes up some aspects of accidents and the magnitude of the accident problem—both in general and in a workplace setting.

Accidents world-wide

From a global perspective, accidents are a major health problem. Each year, there are nearly three million fatalities resulting from accidents or poisoning, of which two million occur in less developed countries (Karolinska Institutet, 1989). According to the same source, injury is the primary cause of death among children and young men in virtually all countries. The medical, social and lost-productivity costs of all injuries are estimated to exceed 500 000 million US dollars each year.

In the USA, an annual total of 4.1 million life-years are lost as a result of accidents and injuries (Committee on Trauma Research, 1985). The corresponding figures for heart disease and cancer are 2.1 and 1.7.

The World Health Organiation (WHO) maintains an international database founded on medical records. Of special interest here are statistics on injuries, since "accidents" are not directly entered into the data set. As seen in the definition below, "accident" corresponds to "unintentional injury", but has a slightly different meaning.

A specialised study of the importance of injuries has been conducted by Krug (1999): "An injury is a bodily lesion at the organic level resulting from acute exposure to energy (mechanical, thermal, electrical, chemical or radiant) interacting with the body in amounts or rates that exceed the threshold of physiological tolerance. In some cases (e.g. in drowning, strangulation or freezing) the injury results from an insufficiency of a vital element. The time between exposure and the appearance of the injury needs to be short." Injuries are often classified as unintentional or intentional. Most traffic injuries, fire-related injuries, falls, and cases of drowning and poisoning are regarded as unintentional. By contrast, homicides, suicides and war-related injuries are categorised as intentional.

It is estimated that 5.8 million people died from injuries world-wide in 1998 (Krug, 1999). This corresponds to a rate of 0.98 per 1000 persons. The death rate for males was almost double (a factor of 1.92) that for females. A conclusion of the study was that injury is the leading cause of death in all age groups. It should be remembered that for every person that dies, several thousands more are injured, many of them permanently disabled.

The magnitude of the problem varies considerably by age, sex, region and income. For example, in the low- and middle-income countries of the Western Pacific the leading injury-related causes of death are road-traffic accidents, drowning and suicide, whereas in Africa they are war, interpersonal violence and traffic. In the high-income countries of the Americas, the leading injury-related cause of death among people aged 15 to 44 years is traffic, whereas in the low- and middle-income countries it is interpersonal violence.

WHO's injury statistics do not identify where injuries occur. This means that the data do not permit comparisons between hazards at work, in traffic, in the home, etc. Information about occupational accidents must come from other sources.

Occupational accidents in the world

Occupational accidents are in themselves a major problem from a world perspective. The International Labour Office (ILO) compiles statistics for occupational accidents and diseases. According to one estimate, 180 000 people a year die from accidents at work, while 110 million are injured (Kliesch, 1988). In a large number of countries, both industrialised and less developed, the frequency of fatal accidents has fallen since the 1960s. For example, it fell—over two decades—by 70% in Japan and Sweden, and by 62% in Finland (Kliesch, 1988). Similarly, the frequency of serious injuries is also falling, at least in industrialised countries. The explanations usually provided for this are that there are fewer people in hazardous occupations and that workplaces have become safer.

These figures are high, but they are also highly uncertain—partly due to missing data. A more recent summary (Takala, 1998) shows still higher figures. For the whole world, the estimated average fatal occupational-accident rate in 1994 was 14 per 100 000 workers. And the total estimated number of fatal occupational accidents was 335 000. Rates differ between individual countries and regions, and also between separate branches of economic activity.

An estimate was also made of the total numbers of deaths related to the workplace (Table 1.1). In total more than 800 000 persons died during 1994. Data and sources of failures have been analysed, and corrections made accordingly. 1.1 million can be considered the best available estimate of annual work-related deaths world-wide. This means that 3000 deaths are caused by work each day (Takala, 1998).

Table 1.1   *Number of work-related fatalities world-wide during 1994 (from Takala, 1998).*

Type of fatality	Number of deaths
Fatalities in workplace	335 000
Fatalities when commuting between work and home	158 000
Fatal occupational diseases	325 000
Total	818 000

A comparison has also been made between a number of countries around the world, as divided into eight major regions (Table 1.2). In total, the size of the world labour force is estimated at 2.7 billion. The inter-country regional fatality rate varies considerably. For the "Established Market Economies" the range in the rate between countries is 1.4–10.

Table 1.2   *Fatal ccupational accidents in the world in eight different main regions during 1994. Fatality rate is given as number of deaths per 100 000 workers (adapted from Takala, 1998).*

Region	Fatality Rate	Number of deaths
Established Market Economies	5.3	19 700
Former Socialist Economies of Europe	11.1	15 600
India	11.0	36 700
China	11.1	68 200
Other Asia and Islands	23.1	80 600
Sub-Saharan Africa	21.0	45 900
Latin America and the Caribbean	13.5	26 400
Middle Eastern Crescent	22.5	41 800
WORLD	14.0	334 900

The incidence of non-fatal accidents has also been estimated. A ratio of 750—between non-fatal and fatal accidents—has been used to provide a foundation for an ILO estimate of non-fatal occupational accidents. The injuries then include 250 million occupational accidents and 160 million cases of occupational disease. These figures are based on relatively conservative estimates (Takala, 1998).

On the costs of accidents

Occupational accidents are also of economic importance—for society, for employing organisations and for the injured persons. At a societal level, the costs are considerable, but difficult to discern and calculate. They are borne by different parts of the health-care system, insurance companies, and so on.

An overview of cost estimates (Dorman, 2000) was recently published by the International Labour Office (ILO). In general, there are many difficulties involved in making such estimates, and it is necessary to make a large number of assumptions. Such studies may be helpful, but should be seen as order-of-magnitude estimates. Estimates have been made for Europe and the USA, but there are no comparable studies of the economic costs of occupational ill-health in the developing world (at any level).

One cited study concerns nine selected European countries (Beatson and Coleman, 1997), which estimates the aggregate economic costs of occupational injury and disease by country. Most costs are in the range 2.5–6% of Gross Domestic Product (GDP).

Another study estimates the economic costs of fatal and non-fatal occupational injuries and illness during 1992 in the USA (Leigh *et al.*, 1996). The total cost was estimated to be 173.9 billion US dollars, corresponding to approximately 3% of US GDP. This was considerably higher than the cost as estimated by Dorman (2000). The greatest cost was related to non-fatal injuries (at 144.6 billion US dollars), while that related to fatal injuries was much lower (at 3.8 billion US dollars). The study also included an estimate of who pays. Based on a number of assumptions, it was concluded that workers bear about 80% of the costs (in one way or another).

The results were summarised by Dorman (2000) as follows:

a) The overall share of occupational injury and illness costs in a typical developed-country economy is substantial, not less than 3% of GDP.
b) Costs may be significantly larger than this, due to the difficulty in identifying the incidence of occupational disease.
c) Workers' compensation plays a significant economic role in determining who bears the costs of disability and premature death.

In general, it can be stated that the total cost of accidents varies considerably between employing organisations. The significance of costs depend on which types of insurance and compensation systems are operated and how sensitive production is to disturbances. Heinrich (1931) and Brody *et al.* (1990) found these costs to be relatively high. But another study (Söderqvist *et al.*, 1990) indicated that the marginal cost of accidents to companies was very low. Reasons for this were that comprehensive insurance policies covered compensation to injured persons, that insurance premiums were independent of the number of accidents, and that there was a certain surplus of personnel.

At a personal level, an accident can create difficulties for the individual in a large number of self-evident ways. There is, however, a long tradition in many countries that people injured at work receive compensatory insurance payments.

Risks at an individual level

People in society engage in a large number of activities, which are hazardous to a greater or lesser extent. The level of risk varies considerably from activity to activity. The ways in which individuals perceive risks and act more or less safely depend on a range of factors. The relations involved have been studied for a long time (e.g. Fischoff *et al.*, 1981) and have become a subject area in their own right. There are explanations related to cultural, economic and other factors, as well as type of risk. In general, higher risks seem to be tolerated in voluntary activities where the individual has a certain degree of control over what is happening. Also, the individual generally obtains some benefit from what he or she is doing.

For many activities, however, risks are not taken on voluntarily; nor does the individual have much control over them. In such cases, higher safety requirements are demanded for risks to be regarded as acceptable. One example is that of air traffic, which is subject to extensive and detailed safety regulation. Another concerns occupational risks, which have to be kept as low as possible.

The estimates in the ILO study (Takala, 1998) give a rate of 14 deaths per 100 000 workers a year, and show that the ratio of non-fatal to fatal accidents is 750. On average, this means that one worker out of 9.5 will be involved in an accident each year.

This figure is high. Even if calculations of average values are uncertain, it should be noted that risks are unevenly distributed in many respects. About half of the working population face a higher than average risk.

Relative frequency can be a rather crude measure of occupational-injury risk. All employees are included in the calculation, but it is almost only those who are directly involved in production who are at risk. For example, data from Sweden indicate that slaughter-house employees run a risk of sustaining an injury that is five times as great as that of workers in general, and 45 times greater than bank or insurance employees.

From the perspective of the individual, perceived accident risk has great importance in terms of welfare. Naturally, awareness of the danger of being killed, disabled or injured has a negative effect on well-being. But, in one sense, experiences of accidents are rare for the individual. Considerable periods of time elapse between occasions of injury. Nevertheless, on average, an industrial worker will sustain an occupational injury two or more times during his or her working life.

Risks at organisation level

Companies and their employees face an extensive and diverse risk panorama. In addition to various commercial risks, a number of specific types can be listed:

- Occupational injuries and health risks.
- Fires and explosions.
- Damage to machinery and equipment.
- Transportation injuries and related damage.
- Product liability and related damage.
- Harm to the environment resulting from the organisation's activities.
- Sabotage.

Large accidents

Historically, the greatest industrial disaster was a chemical accident at Bhopal in India in 1984. The figure for victims varies. Estimates of the number of people killed are between 2000 and 4000; the number of persons sustaining injuries is estimated at between 200 000 and 400 000.

Major accidents, where many people lose their lives receive considerable attention, and are also of great psychological importance. There have been a number of large accidents, associated with places like Chernobyl, Mexico City, Zeebrugge and so on. They all represent events and hazards that are unacceptable to society. There is an extensive literature describing large accidents and their causes (e.g. Cox and Tait, 1998; Jenkins *et al.*, 1991; Perrow, 1984; Reason, 1990). Many lessons can be learned from these accidents. Of particular interest is the role of senior management responsible for planning and operating the installations where the accidents occur.

Severe experiences of this kind have changed and reinforced legislation for chemical and nuclear installations. They have also highlighted the need for safety analysis at major-hazard plants, and shown that it is imperative to include organisational issues in such analysis.

Comment

The prevention of accidents is a world-wide challenge, which has engaged international institutions such as the ILO and WHO. Many different approaches are needed and used. Safety analysis is a useful tool in preventing accidents, if used to a greater extent.

1.2 WHY ANALYSE?

Why perform a safety analysis?

There are a number of reasons for conducting a safety analysis, which can concern either an existing workplace or a design situation. This section offers a summary of what might be relevant to stakeholders of various kinds.

Reduce hazards

The basic aim of a safety analysis is to prevent accidents. In many workplaces the number of accidents is too high, and there is a general need for improvement. In most cases, safety analysis has advantages compared with traditional safety work. A deeper and more systematic analysis will improve understanding of risks, which will better support hazard reduction. A number of examples are given in this book.

Why adopt a specific method?

Is it necessary to use a particular method of safety analysis? Might it not be sufficient to be "systematic"? There are a number of advantages in using a defined method. But this presupposes that a suitable method is chosen according to situation (see chapters 12 and 13); otherwise, utility will be small and perhaps even negative. Using one or several methods of safety analysis may have the following kinds of advantages:

1. A general experience is that far more hazards and ideas for improvements are discovered than in traditional safety work.
2. One part of the explanation for this is that safety analysis offers a complementary perspective and adds to earlier ways of working and thinking.
3. Several methods are based on solid experiences, which have been put together in compact format, with checklists etc. Other ways of obtaining the necessary information would be more time-consuming and difficult.
4. In complex systems, which usually include several hazards, it is essential to work systematically, so that important aspects are not overlooked.

5. Safety in a system depends on co-operation between people in different positions. Using a safety analysis method can give a good format for teamwork because it offers a step-by-step approach. Even if lengthy discussions should arise, and several meetings are needed, it is possible easily to get back on track. This applies especially to the methods described in chapters 5 to 8.

6. In teamwork, application of a safety analysis method can give a more objective touch to discussions. It has been shown to support inclusion of the experiences of workers and operators. It can also be beneficial for discussion of controversial issues.

7. The application of a specified method gives a certain guarantee that safety issues are well handled.

8. The results are documented in a uniform manner.

Stakeholders—different motives

There are several stakeholders who might have an interest in being aware of hazards in a workplace and how to prevent them. But they may have different interests, and also place divergent demands on any analysis. Examples of stakeholders include:

*   Employees, who are usually the persons closest to risk in the workplace and most likely to be injured.
*   People at risk, but not employed. They may, for example, be persons living close to a major-hazard installation, or passengers travelling on a public transport system (which is the workplace of others).
*   Employers and owners of installations (workplaces).
*   Managements of employing organisations.
*   People who produce designs (but are not responsible for operations).
*   Public authorities and legislators.
*   Customers buying and utilising products from workplaces.
*   Insurance companies.

Types of results of a safety analysis

Before discussing the benefits of safety analysis, examples of results are given. Depending on the type of analysis these will vary, and more detailed descriptions are given in later sections of this book. Examples of results from a safety analysis include:

*   An overview of risks in the workplace.
*   A list of hazards in the workplace, where each hazard has been evaluated.
*   A list of recommended safety measures for the workplace.

- A detailed description of how certain accidents can occur, plus an estimation of the likelihood that they will occur and their potential consequences.
- An investigation of an accident showing how technical, human and organisational factors have contributed to the course of events that preceded it.
- A summary of safety features at an installation, and estimations of how efficient they are.
- A better understanding among participants of how production and safety systems function.

Official requirements

Legislation and regulations set norms and define responsibilities for safety and work conditions. There is large variation between legislative requirements and their implementation between countries, but official demands (e.g. CEC, 1989) provide several arguments for conducting systematic safety work. Common elements in legislation are:

- The employer has main responsibility for providing a safe and healthy work situation.
- Safety, health and environment management shall be satisfactorily organised.
- Employees shall be informed about hazards and how to work safely.
- Hazards shall be identified and evaluated, and if necessary reduced or mitigated.

In legislation, terms like "risk assessment" are becoming more common, although they are often used in a rather general sense. Applying safety analysis can be a way of making a systematic and documented assessment of risk.

But there are also more specific regulations that impose demands for risk assessment, one of which applies to machinery for use in the workplace. There is a European Union (EU) directive (CEC, 1989/98), for example, which has been transformed into national legislation in EU states. Its application is supplemented by a EU standard (CEN, 1996).

On sites where large accidents with hazardous chemicals can occur, the official demands for risk management and formal risk assessments are high. In Europe, they are regulated by the Seveso Directive of 1982, later amended (CEC, 1996), and in the United States by the Clean Air Act Amendments (EPA, 1990). For other types of industries, such as nuclear power and off-shore, requirements for systematic safety analysis are high.

The discussion above has concerned specific legislation for workplaces and special types of businesses. But civil and criminal law also offer incentives to prevent accidents and injury, e.g. with regard to avoiding a claim for damages (which in some countries can be very high).

Employees' interest

The most important stakeholders are employees working close to danger, who may be victims of injury if safety arrangements should fail. A good safety analysis can reveal safety problems, and improve both safety and trust in the workplace. An advisable approach in performing such an analysis is to involve representatives of employees. This will improve the analysis, make the most of workplace experiences, and also improve the transparency of the analysis.

Employers' responsibility

Main responsibility for safety and work conditions rests on the employer. In some cases this also applies to the owners of an installation, and any co-ordinator of work performed by people from several companies in a particular workplace. Detailed obligations vary between countries, but this main principle of responsibility is generally valid.

To fulfil such obligations, safety analysis can be an efficient tool in many situations. The records of an analysis can be used to demonstrate that safety has been cared for in a legally adequate manner.

Designers and engineers

One important group of stakeholders consists of designers and engineers. They affect safety in several ways—through technical layout, performance of software, etc. Their responsibility is of key importance, and this can be clarified and highlighted by a safety analysis.

A good example comes from the Engineering Council (1993) in the United Kingdom, which has adopted a *"Code of professional practice on risk issues"* based on ten points. The fourth point reads as follows: "Take a systematic approach to risk issues. Risk management should be an integral part of all aspects of engineering activity. It should be conducted systematically and be auditable. Look for potential hazards, failures, and risks associated with your field of work or work-place, and seek to ensure that they are appropriately addressed."

Safety analysis can be a tool in design and engineering activities. Some reasons for this are as follows:

- It allows requirements, formal and also informal, to be met.
- It creates documentation showing that safety issues are handled correctly.
- The system will be safer.

Professional responsibility

The handling of risks in the workplaces concerns persons in several different occupations, such as managers, designers, and safety specialists. For some, risk issues appear to be peripheral, but for others they are a central part of their job. Safety analysis can be regarded as a tool that should be familiar to specialists and also known by others. But professional responsibility can be interpreted in quite different ways, although the Engineering Council (1993) is very clear about this.

The customer

Sometimes the customer is an essential stakeholder in safety features. The first aspect is the safety of the product the customer buys. This might be a machine, a means for the transport of people, etc. Most types of products are related to a number of safety characteristics and demands; and, for some of these, a documented safety analysis might show that requirements are met.

A number of private and public companies have also included ethical values in their choice of contractors and products, e.g. in the off-shore industry. One such aspect is that contractors should offer a high standard of work environment and safety for their own employees. Systematic safety work provides a way of demonstrating that this is the case.

Economics

There are many connections between economics and safety, all of which would be too complex to describe here. Some case studies, incorporating financial and cost-benefit analyses, are provided in Chapter 15. The economic advantages of applying safety analysis can come from:

- Fewer accidents.
- Fewer production disturbances.
- Systematic identification and elimination of sources of disturbances to production.
- During design, the efficient identification of problems and failures through systematic analysis, the avoidance of such problems, and less need for late and costly corrections to an installation.
- Lower probability of a large accident, fire, etc. (if the analysis had that as a target); financially, this may be reflected in reduced insurance premiums.

Summary of benefits

There are a number of possible benefits associated with using safety analysis as discussed here. The reader may judge how reasonable these are when having gone through this book. The most important arguments for safety analysis are:

- Safety is best improved through the systematic identification and prevention of accident risks.
- Safety analysis documentation can demonstrate that a systematic approach has been used.
- Safety analysis can be good business, especially by preventing production disturbances.

1.3 ON TERMINOLOGY

Introduction

Terms used in relation to accidents, such as hazard, risk, safety, etc., may have different meanings in different contexts. The meanings assigned to them largely depend on traditions that have been established within different academic disciplines and applied in a variety of technical contexts.

Some of the basic terms used in this book are discussed in this chapter. Neither a comprehensive review of the various definitions nor a major search of the literature has been attempted. Some references are provided, but these should be regarded as examples rather than as providing any definitive account of usage.

Accidents and incidents

An *accident* is an undesired event that causes damage or injury. An *incident* is an undesired event that almost caused damage or injury. The term *near-accident* is often used to describe the latter type of event.

One specific term is *major accident*, which usually refers to a large accident in a chemical plant. One formal definition is "an occurrence such as a major emission, fire, or explosion resulting from uncontrolled developments in the course of the operation of any establishment covered by this Directive, and leading to serious danger to human health and/or the environment, immediate or delayed, inside or outside the establishment, and involving one or more dangerous substances" (CEC, 1996).

There are, however, other frequently employed terms. In the medical tradition, the term *injury* is preferred to "accident" (Andersson, 1991). In such a case, "accident" means "an event that results or could result in an injury" (Karolinska Institutet, 1989).

Types of accidents

Some examples of types of accidents are provided below:

A. Accident with direct consequence. A sudden undesired event that is triggered off unintentionally and apparently by chance. The unfavourable consequence is observable within a short period of time. Examples include an accident where someone is crushed in a press, an explosion, and the breakdown of an installation.

B. Accident giving increased probability for injury or damage. This is the same as A, but the consequence is not direct. An example is the increased likelihood of cancer arising from exposure to radiation or chemicals when an accident occurs.

C. Slow deterioration or degeneration. Examples include occupational diseases or environmental destruction caused by continuous exposure or the absorption of repeated small doses of chemical substances, prolonged overexertion, etc.

D. Sabotage. A negative event caused by the wilful action of a person. Sometimes, this type of event is not categorised as an "accident".

The lines of demarcation between these categories can become blurred. For example, the difference between A and C is a question of time. In the case of A, it is a matter of fractions of a second to, perhaps, a number of hours. For C, it is often a matter of years.

This book focuses on sudden, undesired events. Its emphasis is on occupational accidents (those that occur at work), but its philosophy and manner of proceeding are applicable in principle to most risks of types A and B.

Occupational injuries

Occupational injuries can occur in a variety of ways. In general, they can be divided into three categories:

- Occupational accidents—accidents occurring in the workplace.
- Occupational diseases—harmful effects of work that are not due to an accident, such as overexertion injuries, allergies or hearing complaints.
- Commuting accidents—accidents occurring on the way to or from the workplace.

By an occupational accident is meant a sudden and unexpected event that leads to the injury of a human being in the course of his or her work. Generally, the course of events is rapid—lasting seconds or even less. But some, such as those involving toxic gases or cold, might require several hours of exposure before an acute injury is incurred.

In order to make comparisons, some kind of measure of relative frequency is required. The measures vary between countries, often making it difficult for international comparisons to be made. Accident frequency can be given as the number of accidents relative to a given number of people employed, e.g. 1000. One example is the fatality rate, defined as number of fatalities per 100 000 workers during a year. Alternatively, the frequency of accidents can be seen in comparison with number of hours worked, usually 1 million.

Statistical reporting also requires a specification of what is to be counted as an occupational accident. This is often related to insurance-system requirements, and could, for example, be an accident leading to absence from work of three days or more. Other measures are number of days of sick leave per accident and days of sick leave resulting from accidents per employee. One difficulty involved in presenting data in this way is that fatal injuries and those that cause disability have to be allocated a number of equivalent days.

Hazard

The term *hazard* is often used to denote a possible source or cause of an accident. The definition of hazard presented by the International Electrotechnical Commission (IEC) is "source of potential harm or a situation with a potential for harm" (IEC, 1995). "Source of risk" has been proposed as an alternative term (SCRATCH, 1984).

Harm is physical injury or damage to health, property or the environment (IEC, 1995).

Risk

The word *risk* is used in a variety of contexts and in many senses. In general, it can be defined as the possibility of an undesired consequence, but is often regarded as a function of probability and consequence. In everyday speech, its meaning shifts between these two senses. A LARGE risk may refer to the seriousness of the consequences of an event occurring, or the high probability that it will occur, or a combination of the two.

In many contexts, risk is used rather technically. Risk is then a combination of the frequency, or probability, of occurrence and the consequence of a specified hazardous event (IEC, 1995). The term risk may also be used when outcomes are uncertain. But this technical definition also has complications. Vlek and Stallen (1981) have listed definitions of objective risk which are common in the literature:

- Risk is the probability of a loss.
- Risk is the size of the possible loss.
- Risk is a function, generally the product of probability and size of loss.

- Risk is equal to the variance of the probability distribution of all possible consequences of a risky course of action.
- Risk is the semi-variance of the distribution of all consequences, taken over negative consequences only, and with respect to some adopted reference value.
- Risk is a weighted linear combination of the variance of and the expected value of the distribution of all possible consequences.

This is a statistician's view on risk. Perceived risk may be something quite different. According to Brehmer (1987), "the most useful approach to psychological risk may well be to consider risk judgements as intuitive value judgements which express a diffuse negative evaluation of a decision alternative, a general feeling that this is something one does not want."

The diversity of ways in which the concept of risk is used does constitute a problem. In this book, it will be used in its general sense: the possibility of an undesired consequence.

Safety

It is more difficult to define safety. You may say that a thing is safe if it is free from harm or risk, but in practice this state is not obtainable. Safety should rather be seen as a value judgement. A machine or action is regarded as safe if the level of risk of being injured is considered to be acceptable.

This judgement then concerns how large the risk is, what is acceptable, and who shall make this judgement.

In a sense, safety is the opposite of risk, and can be regarded as inversely proportional to the risk (Kumamoto and Henley, 1996). One attempt at definition refers to a "safe system" as one that is free from obvious factors that might lead to injury of a person or damage to property or the surroundings (SCRATCH, 1984).

Safety analysis

There is no broadly agreed definition of safety analysis, but the one employed here runs as follows:

Safety analysis is a systematic procedure for analysing systems to identify and evaluate hazards and safety characteristics.

Definitions of safety analysis and risk analysis are more thoroughly discussed in Section 3.1.

2
Features of systems and accidents

2.1 ELEMENTARY RELIABILITY THEORY

Scope

Reliability theory and probabilistic calculations have great relevance for analysis of systems safety in many applications. However, most of the methods presented in the book are not based on this, instead qualitative approaches dominate. The aim of this section is to provide a short account of reliability theory for readers unfamiliar with the subject. There is a need for a basic knowledge of reliability when making risk assessments, and when safety features are judged.

Issues concerning reliability are also taken up in Chapter 9 on Fault Tree Analysis. For detailed accounts, the reader should refer to the more specialised literature (e.g. O'Connor, 1991; Kumamoto and Henley, 1996; Lees, 1996).

Some concepts

A general definition of reliability is "the probability that an item will perform a required function under stated conditions for a stated period of time".

The probability of survival, or reliability, is usually denoted as R(t), where R stands for Reliability and t is the time period. The failure probability F(t) is the probability that the equipment will break down before the expiry of the time period t. They are related by the formula:

$$F(t) = 1 - R(t) \qquad (2.1)$$

The failure density is the probability of the occurrence of the first failure (after time point 0). It is designated as f(t) and is the negative time derivative of the failure probability.

$$f(t) = -\frac{dR(t)}{dt} \qquad (2.2)$$

Another common variable is the failure rate, sometimes called the hazard rate and often denoted as z(t). It can be said to express disposition to fail as a function of time.

$$z(t) = \frac{f(t)}{R(t)} \qquad\qquad (2.3)$$

Figure 2.1 shows an example of a failure rate function over the lifetime of a system. The failure rate is higher at the beginning, when defective components will fail quickly. These are replaced by components that function normally, resulting in a better-functioning system and a fairly constant failure rate. The increase in the failure rate at the end of the time period is due to the wearing-out of the system. This type of failure rate is known as the "bathtub curve".

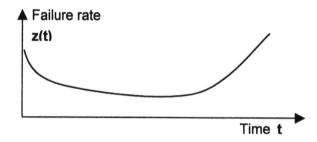

Figure 2.1   The bathtub failure rate curve

Depending on the types of failures that can arise and how the systems are constructed, different statistical models can be applied. Sometimes, the exponential distribution is used. On other occasions, it is the normal distribution or the Weibull distribution. An important case is when the failure rate is constant (time independent), which gives a simple equation for the reliability function:

$$z(t) = \lambda \qquad\qquad (2.4)$$

$$R(t) = \exp(-\lambda\, t) \qquad\qquad (2.5)$$

Other key concepts are Mean Time Between Failures (MTBF) and Mean Time To Failure (MTTF). The two concepts are similar but not identical. MTBF is applied to a population of components or systems where repairs take place. It is a mean value, derived by dividing total operating time by the number of failures. By contrast, MTTF is used for systems that are not repairable.

Series systems

Technical systems usually comprise a number of components. A series system is one that operates only if all components function. Let the various subsystems have reliability probabilities of $R_1(t)$, $R_2(t)$, $R_3(t)$, etc. The reliability for the system as a whole is denoted as $R_K(t)$:

$$R_K(t) = R_1(t) \cdot R_2(t) \cdot R_3(t) \ldots \tag{2.6}$$

If the probabilities of failure are small, an approximate expression for the failure function for the complete system can be obtained from a simple series development:

$$F_K(t) = 1 - R_K(t) = F_1(t) + F_2(t) + F_3(t) \ldots \tag{2.7}$$

Parallel systems

A parallel system is one that fails to operate only if all its components fail to operate. One example is a light fitting with several bulbs; for the fitting to provide no light at all, all the bulbs must fail. The failure function for the system as a whole is given by the equation:

$$F_K(t) = F_1(t) \cdot F_2(t) \cdot F_3(t) \ldots \tag{2.8}$$

Failure data and calculations

The use of reliability techniques gives rise to a need for data on component and system failures, and also information on times taken to make repairs. Data on certain types of human error may also be required. The data can be obtained from data banks or the technical literature, or may be collected directly. Where data are not available, estimates can be made. The level of accuracy required largely depends on the application of the analysis.

Calculations may require the application of advanced mathematical techniques. Moreover, many difficulties arise in obtaining and evaluating the data. One problem is the rapid rate of technological development. New versions of components emerge at such short intervals that the time required to get data about their reliability is not sufficient.

Strategies for improving reliability

There are a variety of strategies that can be applied and combined to improve reliability (e.g. Bergman, 1985). Some examples are:

- *Good design.* Choice of design and proficient engineering work are basic to high reliability.
- *Use of reliable components.* The reliability of a system as a whole depends in part on the reliability of components and subsystems.
- *Maintenance.* Plant maintenance is of decisive importance in terms of reliability.
- *Continuous monitoring.* This applies to specific functions or features of the system. There are many ways of detecting functional deteriorations. For example, the level of vibration or temperature of bearings can be monitored, abnormal values providing an indication that failure may occur. The bearings can then be replaced before breakdown.
- *Regular testing.* (As described below.)
- *Choice of safety margins.* Between load and strength (see below).
- *Redundant systems.* See below.
- *Fail-to-safe* philosophy. See below.

Regular testing

Regular testing of components and system functions can be an effective means of increasing reliability. This requires that test routines are run at regular intervals to investigate whether different subsystems and important components are functioning. This has especially great importance if the existence of a failure is not directly visible. This can give rise to a latent failure, which means that the safety function will not work when required.

Choice of safety margins between load and strength

One aim in design is that strength should exceed load so that breakdown will not occur. Load can take on various forms—weight of objects, pressure of water, tension of electricity, and so on. The concept of "safety margin" is used to describe the relation between size of load and strength of system. Some systems may have a safety margin amounting to a factor of 10, whereas the margin for others is considerably narrower.

In reality, conditions are more complicated. Both load and strength can diverge from their theoretical values, and they should accordingly be regarded as variable quantities. Strength can be reduced for a large number of reasons, such as corrosion, high temperatures in the surrounding environment, etc. A short discussion of how these issues are handled by Fault Tree Analysis is provided in Section 9.4. Otherwise, the reader is referred to the specialised literature (e.g. O'Connor, 1991).

Redundant systems

Where greater reliability or safety is required, redundant functions can be introduced into a system. These can be provided by an extra component (or routine) not needed for the system to operate but which can take over if a defect arises. One example is a battery, which is permanently connected, that takes over in the case of the failure of normal power supplies (see Figure 9.2).

Redundancy can be total, where there are two complete parallel systems. Or, it can be partial, where only the most important components are duplicated. A further distinction can be made between active and passive redundancy. Active redundancy describes situations where the extra system is permanently connected. Passive redundancy refers to cases where the reserve system is activated only when failure occurs, as when an engine-powered generator is started when mains voltage fails.

A further aspect of redundancy concerns whether or not a system is load-bearing. In the case of load-bearing redundancy, the extra component is permanently subject to load.

Fail-to-safe philosophy

It is not possible completely to avoid technical failures. One design philosophy is to construct equipment so that it will always revert to a safe state if a failure occurs. With simple systems, this usually means just that the machine stops. More generally, the creation of such systems requires certain design principles and the selection of critical components that can only fail in a specific manner.

Numerical examples

To illustrate quantitative aspects of different solutions, some simple numerical examples are provided below. These are based on a single module in a system. During a given time interval, the probability of failure is assumed to be 0.02.

To increase reliability, redundancy is achieved by coupling-in an identical module in parallel. The probability of simultaneous failure of both systems is 0.0004 (Equation 2.8). Should a further module be connected in parallel, the probability of overall system failure would fall to 0.000008. Such a drastic reduction, however, is largely only of theoretical value, since it will probably be various types of common cause failures that are the main reason for concern.

Another way of increasing reliability is to reduce the time interval between tests. If tests are carried out ten times as frequently, the original probability of failure might be reduced to approximately 0.002. It is possible, however, that testing introduces new sources of failure, which will reduce the scale of improvement.

Common cause failures

In theory, a high level of reliability can be achieved—as in the example above. When calculating reliability or failure rates, it is often assumed that component failures are independent. But the presence of common cause failures can drastically reduce reliability.

A common cause failure can arise from a fault in the manufacture of several modules, which would all give them the same defect. Also, several components might be exposed to an unsuitable environment, such as high temperature, thus leading to a deterioration in product life.

Taking account of the possibility of common cause failure is particularly important in the case of redundant (parallel) systems.

Some comments

A solution involving redundant modules in a safety system increases the probability that it will function. The system also becomes more complex, both to manufacture and to maintain, which means that costs rise. High reliability has a price.

2.2 ON HUMAN ERROR

Introduction

"To err is human", is an old proverb. It is nearly always the case that human error lies behind an accident. Such errors can be of many different types. They may be simple, such as when someone hits his thumb. They may also be advanced cognitive errors, as when an important safety system is designed in the wrong way. All people make unintentional mistakes. Usually, it is only when they have unfavourable consequences that they get attention.

The aim of this section is to provide background theoretical knowledge on human errors and human behaviours. In performing safety analysis, such issues are important in several ways:

1. At the identification stage of an analysis, human error often comes up as an essential element.
2. In the design of safety improvements, a better understanding of human behaviour can provide for more efficient solutions.

The importance of human action to system safety has received increasing attention over a number of years, and there is now an extensive literature on the subject. This section offers a short discussion of the issues involved. Otherwise, the technical literature and a number of reviews can be referred to (e.g. Petersen, 1982; Hale and Glendon, 1987; Reason, 1990; Hollnagel,

1993). Brief overviews of methods related to human errors are also considered in sections 10.3 and 10.4.

The human factor

The reader should note that the term, the "human factor", should not be confused with the wide usage of "human factors" in the context of ergonomics. In fact, the study of "human factors" is often considered to be equivalent in meaning to "ergonomics" and, in much of the modern literature, the term has acquired a very wide sense. One definition (Health and Safety Executive, 1989) runs as follows: "The term human factors is used here to cover a range of issues. These include the perceptual, mental and physical capabilities of people and the interactions of individuals with their job and working environments, the influence of equipment and system design on human performance, and above all, the organisational characteristics which influence safety related behaviour at work."

A popular reaction to an accident is to blame it on the "human factor". A newspaper usually accepts this as the main explanation. Often, it is a representative of an authority or a safety manager who couches the explanation in these terms. This is not wrong in itself. All accidents are related to human actions. People use a piece of equipment, and people also make decisions on how the equipment is designed, and how work is planned. *But, there is a certain "ring" to the term, the "human factor". It implies that accidents are due to irrational and unpredictable elements in a situation, and that nothing can be done about them.*

Moreover, it is often the person receiving an injury who is regarded as the "factor". It can sometimes be related to a scapegoat thinking, and an attitude to put the blame somewhere else. Such attitudes easily lead to passivity and come to be an obstacle to accident prevention.

Accident proneness

One early attempt to systematise knowledge on accidents involved focusing on the role of the individual. This resulted in the theory of accident proneness, devised by Farmer and Chambers (1926). They put forward the hypothesis that it was certain categories of people who were most likely to be the victims of accidents.

According to this theory, some individuals possess certain stable attributes that make them particularly liable to accidents. A consequence of the theory is that accidents should be combated by selecting individuals—e.g. by using various testing procedures, and allocating them tasks which are appropriate. However, it has proved difficult to distinguish attributes of the individual from variations in exposure to hazards in the work environment. The theory does not provide a fruitful general explanatory model. But, in certain contexts, where stringent demands are placed on the individual to act safely, it has served to provide a basis for the selection of job tasks (e.g. for airline pilots).

Why is it that more accidents do not occur?

The problem of accidents can also be approached from the opposite direction. From looking at a construction site, where people are climbing from place to place, where there is a lot of traffic, etc., it might seem that an accident must occur every day. Severe physical hazards do give rise to a higher frequency of accidents, but not one that is extremely high. One explanation lies in risk compensation. Construction workers adapt their behaviours in the light of the risks they face. In general, this question concerns safe and unsafe behaviour at work and how risks are perceived in the workplace (e.g. Hale and Glendon, 1987).

Human reliability

People make mistakes, but more often they do things right. Instead of focusing on human error, an alternative starting-point might be to regard the person as a safety resource rather than a hazard. Even though this might appear to be a philosophical remark, this approach can provide a different basis on which systems can be designed.

An example
A man had worked for a long time with a packaging machine. After ten years he received a severe crush injury from the machine. No special circumstances were found to apply and the accident was treated as the result of "inexplicable error"; perhaps he had been tired and acted clumsily.

In his daily work, he had had to correct a disturbance to production roughly five times a day. This was dangerous if he made a mistake. On one occasion, he did make a mistake, and was injured as a result (after ten years, which is the average time between accidents per person). This means that he had managed to accomplish the task successfully 10 000 times without being injured. Should he not really be regarded as reliable?

At the packaging machine, the man corrected mistakes that were made elsewhere in the plant. One might ask about the defects that caused trouble so often. Why had these not been discovered and corrected? If there was no remedy, why were safer routines not used when he corrected the disturbance?

If human beings are regarded as the cause of occupational hazards, safety strategies might be based either on removing people from production through automation or on the strict supervision of work. A different approach would involve pointing to the role of the human being as a problem solver and safety factor in technical systems. Such an approach raises a number of questions, e.g. on the skills of operators and on the needs for information on the technical system, for organisational support and for the feeding-back of previous experiences.

Human reliability has also obtained a more precise technical meaning. It can be defined as the probability that a job will be successfully completed within a required minimum time (Embrey, 1994). Methods of "Human Reliability Assessment" are discussed in Section 10.3.

Perspectives on human error

Embrey (1994) has proposed a summary incorporating four different perspectives on human error:

a. Traditional safety engineering focuses on the individual rather than systemic causes of error. The basic assumption is that the individual has a choice whether or not to behave in an unsafe manner. The implication is that the responsibility for accident prevention ultimately rests with the individual worker.
b. Human-factors engineering and ergonomics see errors as a consequence of a mismatch between the demands of a task and the physical/mental capabilities of an individual or operating team.
c. The cognitive-engineering approach emphasises that people impose meaning on the information they receive, and that their actions are almost always directed at achieving some explicit or implicit goal.
d. The sociotechnical-systems approach considers the impact of management policy and organisational culture on the individual's behaviour.

On different types of human error

The incidence of human error varies considerably, and it differs between individuals. Moreover, the proneness of the individual to err varies with time and situation. This can be due to a large number of factors, both internal and external to the individual (e.g. Petersen, 1982).

There can be very different types of errors, and they can be defined and classified in different ways. One working definition has been suggested by Reason (1990). A slightly simplified version of his account is provided below.

Human error is "taken as a generic term to encompass all those occasions on which a planned sequence of mental or physical activities fails to achieve its intended outcome, and when these failures cannot be attributed to the intervention of some chance agency".

Slips and lapses are "errors which result from some failure in the execution and/or storage stage of an action sequence, regardless of whether or not the plan which guided them was adequate to achieve its objective".

Slips occur when an action does not go as planned, and they are potentially observable, e.g. slips of performance or slips of the tongue. The term lapse is often used to refer to "more covert error forms, largely involving failures of memory, that do not necessarily have to manifest themselves in actual behaviour and may only be apparent to the person who experiences them".

Mistakes can be defined as deficiencies or failures in the process of making judgements or inferences. Mistakes are complex and less well understood than slips. This means that they generally constitute a greater danger, and they are also harder to detect (Reason, 1990).

On models and explanations

From the end of the 19th century and onwards, many have sought to understand why people make errors in their thinking and in the performance of actions. Reason (1990) has provided an interesting review of developments over the last hundred years.

The most renowned of the pioneers was Sigmund Freud (1914) who found meaning in what were apparently random and day-to-day slips and lapses. Analysis of the errors often permitted the detection of explanations in unconscious thought processes, which had their origins in psychological conflicts.

In cognitive psychology, the idea of "schemata" plays a central role. The term was first adopted by Bartlett (1932). He presented the view that schemata were unconscious mental structures composed of old knowledge, and that the long-term memory comprised active knowledge structures rather than passive experiences.

According to Reason (1990), "the current view of schemata is that they constitute the higher-order, generic cognitive structures that underlie all aspects of human knowledge and skill. Although their processing lies beyond the direct reach of awareness, their products—words, images, feelings and actions—are available to consciousness. The very rapid handling of information in human cognition is possible because the regularities of the world, as well as our routine dealings with them, have been represented internally as schemata. The price we pay is that perceptions, memories, thoughts and actions have a tendency to err in the direction of the familiar and the expected."

One model to which reference is often made is based on distinguishing between three different performance levels (Rasmussen and Jensen, 1974; Rasmussen, 1980).

1. *On a skill-based level* people have routine tasks with which they are familiar and which are accomplished through actions that are fairly direct. The errors have the nature of slips or lapses.

2. *On a rule-based level* people get to grips with problems with which they are fairly familiar. The solutions are based on rules of the IF/THEN type. A typical type of error occurs when the person misjudges the situation and applies the rule incorrectly.

3. *On a knowledge-based level* people find themselves in a new situation where the old rules do not apply. They have to find a solution using the knowledge that is available to them. On this level, errors are far more complex by nature, and may depend on incomplete or incorrect information, or limited resources (in a number of different senses). Rasmussen has suggested that problem solution involves eight steps: activation, observation, identification, interpretation, evaluation, goal selection, procedure selection and activation. He does not assert that each step is taken in this particular order. The decision-maker can jump between steps in order to attain a solution to a problem.

Decision-making can be regarded as a conscious and logical process during which costs and benefits are weighed against each other. Such an account presupposes that the alternatives are relatively clear and that a sufficient amount of definite information is available. In more complex situations, the limitations of people themselves mean that this is not a particularly accurate model. "The capacity of the human mind for formulating and solving complex problems is very small compared with the size of the problems whose solution is required for objectively rational behaviour in the real world—or even for a reasonable approximation of such objective rationality" (Simon, 1957). "The limitation in human information processing gives a tendency for people to settle for satisfactory rather than optimal courses of action" (Reason, 1990).

Violations

Violations represent a further type of human error. By a violation is meant an intentional action which is in breach of regulations, either written or oral. The intention, however, is not to damage the system. Deliberate intention to harm is better described as sabotage. It is difficult to draw a sharp dividing line between errors and violations, and perhaps this is not necessary. In some cases, conscious deviation from what is accepted practice may be seen as a deviation. In many situations, this also applies to risk-taking. Violations can of course be committed both by people who work directly with a piece of equipment and those involved in planning and design.

There are many reasons why people act in breach of regulations. Some examples:

1. The person does not know that the action constitutes a violation. He or she may not be aware of the regulation, or may not be conscious that the action in question represents an infringement.
2. The person is aware of the regulation, but forgets it, e.g. if it seldom applies.
3. The regulation is perceived as unimportant, either by the person himself or by those around him.
4. There is conflict between the regulation and other goals.
5. The regulation is thought to be wrong or inappropriate, either with or without reason.

Risk-taking

On an individual level, accidents can be related to risk-taking, i.e. actions are taken which are known to be dangerous and may also be forbidden. Risk-taking has many similarities with committing a violation. It is unknown how great the problem of risk-taking actually is. The accident rate for men is higher than that for women, and, on average, young men run a greater risk of being injured than older men. Part of the explanation for this may lie in different propensities to take risks, but it is probably also the case that men who are young undertake tasks that are relatively more hazardous.

At the same time, it often pays to take a risk. Hazardous ways of working can be faster and less strenuous, and thereby give rise to higher productivity. In the author's experience, over-ambition at work is a common explanation for the taking of risks. Even in risky jobs, accidents are relatively uncommon, happening to a person perhaps once every ten years. This is why it is generally the benefits of risk-taking that are most visible, encouraging people to take risks and easily making risky behaviour habitual.

In general, piece work reinforces risk-taking. In forestry work in Sweden in 1975, a piece rate system was replaced by a fixed form of remuneration. The frequency of accidents has fallen by around 30% (Sundström-Frisk, 1984). The taking down of already-felled but suspended trees is the job task where the benefits of risk-taking are most apparent. For this task, the number of accidents fell by 70%.

Risk-taking is usually thought of in relation to workers. But the issue is also highly relevant for persons at higher organisational levels. Slightly different mechanisms are valid here; they concern lack of time, ignorance of hazards and responsibilities, and lack of commitment. The major difference is that people other than themselves face the risks.

Safer behaviour

Human error and risky behaviour are strongly affected by technical design and organisational structure, and also by social patterns in the workplace. There are many different ways of getting individuals to behave more safely, which have a greater or lesser degree of success. Information campaigns to improve safety are quite common, but they tend to have only short-term and marginal effects (e.g. Saari, 1990).

Probably the most common way of attempting to promote safer behaviour is to introduce stricter rules, with supervision to ensure that they are followed. Attention is drawn to a type of erroneous behaviour, and an attempt is made to correct it. However, introducing rules that will achieve results is difficult, and often requires a lot of thought (e.g. Hale, 1990).

There have been a number of experiments with "performance feedback", the goals of which are to get workers to make greater use of personal protective devices, employ safer working methods, and so on. The feedback, for example, may consist of someone noting down the proportion of workers using a particular piece of safety equipment and reporting results in the form of a diagram that is displayed where everyone can see it. Most studies have shown improvements in behaviour, but the effect on accidents has been followed rather seldom (McAfee and Winn, 1989; Saari, 1990).

One pre-condition for either way of proceeding to succeed is that the technical and organisational conditions are right. Another is that correct and safe ways of working can be specified.

Tolerant and forgiving systems

One important property of technical systems concerns how important it is for the operator to always take the correct action. If a simple error directly gives rise to an acute hazard, the system should be regarded as dangerous. Systems can be "forgiving" to a greater or lesser extent. Such forgiveness might involve providing the operator with an indication that an error has been committed, before, for example, a machine movement is triggered off. Or, opportunities might be made available to correct an error, so that the error in itself does not have serious consequences.

The individual and safer ways of working

The conditions for the individual to work safely can be expressed in terms of three key words, *Know*, *Can* and *Will* (Bird and Loftus, 1976). Let us take the operator of an automated machine as an example. The safety of his or her work depends on a number of factors. Some of these are as follows:

- **Know**—that he knows how to work safely.

This depends on the training he has received for the job task in question. For example, machine manuals need to be designed so that they are comprehensible. Knowledge is needed on functional properties, both when the machine is in normal operation and when disturbances of different types occur.

- **Can**—that it is possible to work safely.

For example, it should be easy to stop the machine, and also re-start it without having to go through a complicated procedure. Safety devices should be designed so that they are not a hindrance to doing the work. In operating a computer-controlled machine, the interface should be designed so that the operator understands how to act correctly and can avoid mistakes.

- **Will**—that he has the motivation to work safely.

The motivational condition is in many ways the hardest to satisfy, especially from a long-term perspective. It is essential that the operator is aware of the hazards, and that conditions promoting risk-taking are kept to a minimum.

2.3 SYSTEMS AND ACCIDENTS

Scope

The reasoning in this book is largely based on a systems perspective, which includes three major dimensions:

- The production system (to be analysed).
- Risks, how accidents can occur.
- Safety, how accident risks can be controlled and reduced.

These dimensions relate to any method of safety analysis and also (more or less explicitly) to the theoretical model underlying the approach. An understanding of these aspects is essential in choosing and applying a method of safety analysis. The aim of this section is briefly to present these dimensions, to which we will return in a discussion of theoretical models (Chapter 14).

The production system in general

A production system can be seen as a number of elements that must interact for a desired result to be achieved—and also for avoiding accidents. The main components are:

1. Technical equipment and physical conditions.
2. Individuals within the company.
3. Organisation and activities.
4. Surroundings, including society.

Another dimension of the systems approach concerns the life cycle of the production system, and all the different states that can occur. Safety considerations should apply during planning, design, production start, operation and decommissioning. The operational phase includes both normal and disrupted production, maintenance, and system change.

The types of technical equipment that may be involved in accidents vary considerably, ranging from hand-held knives to the advanced computer systems that now control production plants.

A fundamental aspect of safety concerns organisation. This will guide how machines are designed and maintained, how job procedures are planned and supervised, etc. There are a wide variety of situations, all of which affect safety in different ways.

Explanations for accidents

Explanations for accidents show large variation, and there is no uniform, universally applicable theory. But in several methods of safety analysis there is a clear model of how accidents occur.

It is common to look for the "the cause"—the accident then being regarded as the product of one event and as having just one explanation. The drawback with simple explanations, which treat just a selected portion of reality, is that they may prevent problems from being solved effectively.

For this reason, explanations and theories are useful, especially when they provide sufficient insight into why accidents occur and how they can be prevented.

Accident models

There are a large number of different models available. For example, there is a long tradition of research into individual behaviour and individual characteristics. On the medical side, there are a variety of different epidemiological models. By studying the interaction between the individual (the injury's "agent") and the environment, as has been done in the case of infectious diseases, information on causes can be obtained (Gordon, 1949).

The so-called "Domino Theory" was first launched by Heinrich (1931) in the 1930s, and has had great significance for practical safety work over many years. An accident is described in terms of a sequence of events, unsafe acts and physical hazards. If these elements can be eliminated, accidents can be prevented. However, the Domino Theory has also been subjected to criticism (e.g. Petersen, 1982) largely on the ground that it describes accidents in far too simple a manner. Nor does it explain why unsafe actions are taken, or why mechanical or physical hazards arise.

There are a number of models that are system-oriented to a greater or lesser extent (for an overview see Kjellén, 2000). Simply expressed, many of these models view the company as a system, with technical, human and organisational resources that have to interact for certain results to be obtained. Further, they are based on the view that there are always contributory causes of an accident. These are regarded as abnormal system effects, and might be due either to the failure of individual components of the system (including human beings) or disruption to the interaction between them.

Of special interest are the lessons to be learned from accidents with major consequences, which are related to large organisations. Such accidents show complicated patterns of organisational failures and ways of controlling hazards. The systems tend to have advanced safety features, and the theoretical probability of an accident often appears to be extremely low. There are several explanations of why safety systems fail (Perrow, 1984; Reason, 1990, 1997). Examples are that safety systems do not cover all eventualities, that the systems do not function as planned, or that they deteriorate over time.

Perspectives on accidents

An illustration of how explanations of accidents have changed over time is presented in Figure 2.2. Ever greater attention is being paid to organisational issues, at the same time as production systems are conceived as more complex.

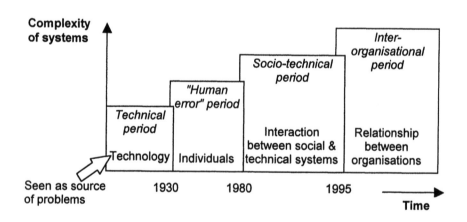

Figure 2.2   A long-term perspective on sources of accidents (adapted from Wilpert and Fahlbruch, 1998).

A somewhat different perspective is sketched out in Figure 2.3. To the left, the cause of an accident is treated as a random failure, which could be a human error or failing component. The accident is seen as a random event subject to little control by the company.

To the right, by contrast, the explanation for the accident is seen as the result of a failing company management system. Control exercised by the company is not sufficiently effective to avoid a large accident. The explanation also comes up to a higher systems level. In between these extremes, there are a number of intermediary forms.

The first explanation might be suitable in small systems working more or less independently. The second is more systems oriented, and becomes especially essential when large potentially dangerous systems are scrutinised.

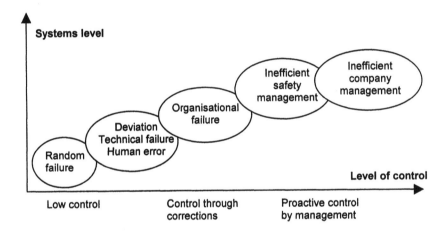

Figure 2.3   Explanations for accidents in relation to level of control at a company.

Safety features

There are many ways of avoiding accidents. Technical and organisational solutions can be applied, according to situation. In some of the procedures described below, there are specific methodologies for finding safety solutions (especially in the cases of energy, deviation and safety-function analysis).

At company level there are often safety management systems, about which a broad range of literature has been developed. Safety analysis can be seen as one of several tools to be applied in this context.

A discussion of functions and methods for analysing certain safety aspects can be found in Chapter 10.

3
Safety analysis

3.1 WHAT IS SAFETY ANALYSIS?

General

Analyses of risks are conducted within a variety of professional areas, and in various ways. This means that the meanings of a number of concepts also vary. There are standards that define parts of the terminology for certain application areas. This section will define risk and safety analysis and also present some alternative definitions.

Definition of safety analysis

There is no broadly agreed definition of safety analysis. The one proposed here is:

> *Safety analysis is a systematic procedure for analysing systems to identify and evaluate hazards and safety characteristics.*

This definition is wide, and it includes both qualitative and quantitave methods. It also covers the more specific definition of risk analysis below. This book is generally concerned with qualitative analyses, although some quantitative aspects are discussed.

In most of the applications presented in this book, an essential part of the analysis is the generation of proposals for improving safety. A common aim is to obtain an overall picture of hazards within a system.

Definition of risk analysis

Within the area of dependability and reliability there is an international standard (IEC, 1995) that defines "risk analysis" and a number of related terms. According to this standard:

> *Risk analysis is the systematic use of available information to identify hazards and to estimate the risk to individuals or populations, property or the environment.*

In the standard, risk is defined as a combination of the frequency, or probability, of occurrence and the consequence of a specified hazardous event. Risk analysis is also sometimes referred to as probabilistic safety analysis (PSA), probabilistic risk analysis (PRA), quantitative safety analysis, and quantitative risk analysis (QRA).

Terminological differences

Other definitions are also applied, which sometimes causes confusion. Terms are used differently according to application area. In the chemical industry, the preferred term is risk analysis for all type of methods. In the nuclear industry, safety analysis appears to be more common. Examples of other common expressions are "risk assessment" and "hazard assessment".

It is good to be aware of the variety of terms, and that different meanings might be entailed.

The systematic approach

One of the keywords in the definition of safety analysis presented above is "systematic". If an analysis is to be of good quality, it is essential to consider the points below.

Let us suppose that a particular production system is to be analysed. The analysis might apply to an existing installation or to production facilities that are still at the planning stage. There are several different aspects to a systematic approach:

- A general procedure for the analysis is defined.
- Gathering of information on the system provides the basis for the analysis and must be carried out systematically.
- The entire system and the activities within it should be included in the analysis. The analysis needs to be designed so that important elements are not overlooked. A main thread must be identified and followed.
- A systematic specified methodology is required for the identification of hazards.
- The risks to which these hazards give rise need to be assessed in a consistent manner.
- A systematic approach is required when safety proposals are to be generated and evaluated.

The systematisation of experience

A method for safety analysis can also be seen as a compressed account of previous experiences. For example, the checklists used in several methods represent summaries of what has previously been found to be important in terms of the identification of hazards.

The development of analyses and safety activities are much "accident-driven". Or, as Reason (1990) puts it, "events drive fashions". People have been forced to rethink, in one way or another, by their own experiences. Perspectives on accidents and strategies for the analysis of risks have been governed to a considerable extent by accidents that have already occurred.

3.2 SAFETY ANALYSIS PROCEDURE

Steps in a safety analysis

A safety analysis consists of a number of co-ordinated steps, which jointly make up a procedure. Figure 3.1 presents one example of a safety analysis procedure. A consistent theme of this book is that we assume that the aim of an analysis is to achieve a reduction in the level of risk.

For this reason, decisions on and the implementation of safety measures have been shown. It should be remembered that there are a variety of other flow charts, used by various authors, which have different areas of application.

The three central elements in the figure are the identification of hazards, the assessment of risks, and the making of proposals for safety measures. The form that these activities take is related to the method employed, while the other elements are of a more general nature.

Introductory part of the analysis

PLAN
The planning of analyses is extensively discussed in Chapter 13. One of the first steps is to take the decision to conduct a safety analysis. This involves consideration of:

- What is to be analysed, what limits to the analysis are to be set, and what assumptions are to be made.
- The aim of the analysis. This might be finding ways to increase the level of safety, or a general evaluation of safety. In the latter case, the stage "Proposals for safety measures" disappears from the analysis.
- Choice of methods and manner of approach.

GATHER INFORMATION
Information on the system to be analysed is needed. This applies to its technical design, how the system functions, and which activities are undertaken. To a great extent, the need for information is governed by the choice of methods to be employed.

Other useful information may concern accidents that have occurred, near-accidents and disturbances to production. If probabilistic analyses are to be conducted, data on frequencies of failure for the components used in the system are also needed.

In the cases of analyses of installations that have been in operation for some time, information is relatively easily accessible. When the analysis concerns production facilities that are still at the planning stage, it is more difficult. Information can then be obtained from drawings, written and oral descriptions, and from experiences of similar installations.

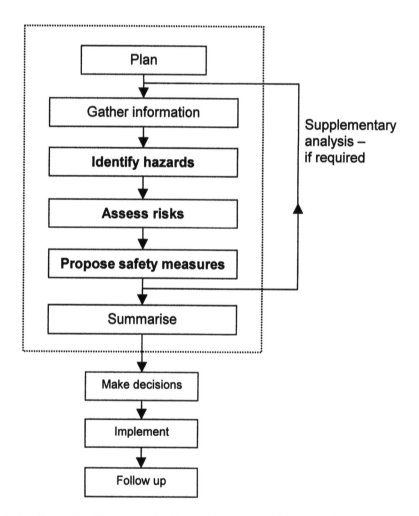

Figure 3.1   Example of stages of safety analysis procedure, and how results are used.

Central parts of safety analysis

IDENTIFY HAZARDS

The central component of most safety analyses is the identification of hazards and other factors in the system that might lead to accidents. One aim should be to discover the major sources of danger and which factors might trigger off an accident. The method selected determines how the process of hazard identification proceeds. When a specialised method is used, certain types of hazards are discovered, but others may be overlooked.

ASSESS RISKS

An assessment is made of risks in the system. Such assessments can take different forms (as discussed in detail in Chapter 4). One application of risk assessment is to judge whether a system is safe enough, or if safety measures are necessary.

In a quantitative analysis, values for probabilities and consequences are estimated. In the case of qualitative analysis an evaluation is made without numeric values.

PROPOSE SAFETY MEASURES

If needed, risks can be reduced through one or several safety measures. The reduction can apply to either consequences or the probability that such negative events will occur. Some of the methods include a systematic procedure for the identification of potential safety measures.

SUPPLEMENTARY ANALYSIS

In the course of conducting the analysis, it might be discovered that more detailed examinations are required, or that a supplementary method is appropriate.

SUMMARISE

The results of an analysis are summarised so as to provide a basis for decision-making. The summary might include a list of the hazards observed, proposals for safety measures, and an account of the assumptions and conditions under which the analysis was conducted. This will finish the analysis.

After the analysis

Make decisions
We assume that the summary is then used as a basis for decision-making. Usually this is not a part of the analysis, and decisions are made somewhere else in the company.

Implement safety measures
For an analysis to have an effect, the safety measures decided upon must, of course, be implemented.

Follow up the analysis
It is also a good idea to make plans to follow up the analysis. This can involve making certain checks on the analysis, establishing that measures have been implemented, and examining results at a later date. For example: Has the number of accidents fallen? How has production been affected?

3.3 A SHORT METHODOLOGICAL OVERVIEW

Choice of methods

There are many different methods of safety analysis (for further discussion, see Chapter 12). Clearly, anyone who is to work with safety analysis must make a choice between the large number of methods available. In general, it can be said that any one specific method will only cover a limited part of the risk panorama.

Some of the criteria that might be used in choice of method are:

- That the method provides the support necessary to sustain a systematic approach (as discussed above in Section 3.1).
- That the method is easy to understand and apply.
- That an analysis can be conducted even when information on the system is incomplete. For example, this may be the case when an analysis is conducted of plant or equipment that is still at the planning stage. This may give rise to poorer accuracy, but the analysis is still worthwhile.
- That an analysis can be conducted with a reasonable amount of effort, taking anything from part of a day to one or several weeks.

This book highlights a set of around ten methods. It is based on the author's own selection. The selection was guided by several considerations, in particular that the four criteria above should be covered, and that a range of complementary approaches is presented.

Overview of methods

Table 3.1 provides a sample of the methods presented. A more extensive overview of a number of methods and a comparison between them are presented in chapters 11 and 12. The three first methods have mainly a technical perspective in describing the system and explaining the cause of an accident.

The other four methods in the table have a more systems oriented perspective. To a greater or lesser extent, they consider the connections between technical aspects, people in the system, and the organisation.

The order of the methods presented in Table 3.1 is different from that in the book. The principle for the body of the book is to start with the easiest methods (Energy Analysis and Job Safety Analysis), and then proceed onto more advanced techniques.

Table 3.1   Some methods of safety analysis (chapter or section reference in brackets).

Method	Comments
Technically oriented	
Energy Analysis (5)	Identifies hazardous forms of energy
Hazard and Operability Studies (HAZOP) (8)	Identifies hazardous deviations in chemical process installations
Fault Tree Analysis (9)	Logical model of the causes of an accident
Systems oriented	Technical, human and organisational factors
Job Safety Analysis (6)	Hazards in work tasks
Deviation Analysis (7)	Identifies hazardous deviations in equipment and activities
Task Analysis (11.4)	Analysis of tasks of people
Safety Function Analysis (10)	Analysis of the safety characteristics of a system

The analytical procedure

Four of the methods have a similar analytical procedure. They are Deviation Analysis, Energy Analysis, HAZOP, and Job Safety Analysis. The different steps are taken in a planned sequence. This facilitates undertaking the analysis, and also makes it easier to plan. The key steps that these methods have in common are as follows:

1. A system is divided into several components, which involves the construction of a simplified model of the system. This step is called "*structuring*".
2. For each component of the system, sources of risk (hazards) or other factors related to the risk of accidents are identified.
3. Some form of risk assessment is carried out.
4. In most cases, a stage at which safety measures are proposed is included.

Quick analyses

To obtain a quick overview of hazards at a plant or from a piece of equipment, some type of rough analysis can be conducted. Such an analysis represents a compromise between thorough analysis and unsystematic observations (see Section 11.7).

4

Risk assessment

4.1 INTRODUCTION

Scope

In most cases, risk assessment forms an important part of a safety analysis. The seriousness of an identified hazard needs to be evaluated. In some of the methods described, risk assessment constitutes a specific stage in the analytical procedure.

As mentioned in Chapter 1, the literature on safety analysis contains a variety of terms, the interpretations of which can vary. Also, risk assessment has a number of meanings depending on the application and the kind of problem addressed. A somewhat simplified account is discussed below. The issue of assessment is sometimes complicated, and has been much debated. Some parts of this discussion are summarised in Chapter 14.

In this book, safety analysis is treated as having three principal components:

- Hazard identification (identification of sources of risk).
- Risk assessment.
- Generation of safety proposals.

Aims of risk assessment

The general aim of a risk assessment is to provide a basis for deciding whether a system is acceptable as it is, or whether changes are necessary. A further purpose is to distinguish between important risks and less important ones.

Some examples of more detailed objectives are given below. They do not exclude each other and are usually determined by the general goal of the safety analysis.

- Give an estimate of the "size" of the risk.
- Approve the system by comparing the risk level with given criteria.
- Judge whether system improvements are needed to increase safety.
- Provide a basis for alert, e.g. establish whether it is essential to assure that the safety system is not degrading the safety properties.

Subject of assessment

How an assessment is approached also depends on what is to be assessed. At one extreme, it might be an entire plant and its overall hazards to employees and the public. This can apply to certain chemical plants, offshore platforms, and nuclear power plants. These kinds of installations are usually thoroughly regulated by authorities that place high demands on performing comprehensive safety analyses, e.g. on the basis of the Seveso Directive (CEC, 1996).

In many of the methods presented here, the outcome of the identification stage is a list of sources of risk (hazards, possible causes of an accident, etc.). Each of the items on the list needs to be assessed individually. The number of items may be quite high, sometimes up to several hundred.

This chapter focuses on types of risk assessments that are relatively simple and concrete by nature.

Criteria and norms

In principle, assessment is made against some kind of norm for what is required. However, the availability of clear and unambiguous norms is the exception rather than the rule.

The directives issued by the authorities offer one basis on which accident risks can be assessed. These, however, are mainly general by nature and do not cover all types of hazards. In some situations and for certain types of equipment, fairly concrete information can be obtained on whether or not a risk is acceptable. But there are also many formulations of the type "Protection against injury shall be adequate" or "that risks should be As Low As Reasonably Achievable (ALARA)". To establish what is "adequate" or "reasonable" remains a matter of judgement.

Types of risk assessments

With a little simplification, risk assessments can be divided up into four main groups:

1. Informal risk assessments.
2. Quantitative assessments based on estimates of consequences and probability (discussed in Section 4.2).
3. Qualitative assessments (discussed in sections 4.3 and 4.4).
4. Safety integrity assessment

As a complement to risk-based approaches, one category of assessments is concerned with assurance of the safety integrity of a system. Then, the aim is to evaluate the adequacy of barriers and safety functions in the system. Some methods of safety analysis are based on this approach (Chapter 10).

Informal assessments

In this context, an informal risk assessment is one that is not a planned part of the analytical procedure and not based on any specific documentation of risk. It takes the form of a general statement on the level of risk.

A number of informal risk assessments are made in connection with the identification of hazards. In practice this can lead to a hazard not being included on the record sheet. This might be because the probability of the occurrence of the event seems low or that its consequences seem minor. Or there may be other justifications, of a greater or lesser degree of validity.

In practice, such omissions cannot be avoided, as the number of identified hazards may be large. On the other hand, it is not acceptable that hazards are dismissed on the grounds that they are "just part of the job", etc. Thus, an awareness of the problems involved in risk assessment is needed even at the hazard-identification stage.

4.2 QUANTITATIVE ASSESSMENTS

Principles

A quantitative approach to risk assessment is used in many applications of safety analysis. The probability that a certain accident will occur and the scale of its consequences are calculated or estimated. The quantitative measure of risk can then be utilised to judge whether or not a hazard is acceptable.

This procedure is often referred to as probabilistic safety analysis or probabilistic risk analysis (IEC, 1995). In such applications, risk assessment has two major components:

- Risk estimation (making estimates of probabilities and consequences).
- Risk evaluation (making an overall judgement of the risk, e.g. in terms of its acceptability and how it is perceived).

Risk evaluation presupposes some kind of criteria or acceptance limits (for a particular risk). Figure 4.1 illustrates relations between frequency of occurrence and the size of consequences and limits of acceptance.

Hazard A, for example, has both a low frequency (probability of occurrence) and a small consequence if an accident should occur. The risk is acceptable, and is below the limits of acceptance.

Hazard C has a high frequency and large consequence, and is above the limit of what is unacceptable. Something needs to be done to reduce consequence and/or probability, if the analysed system shall be approved.

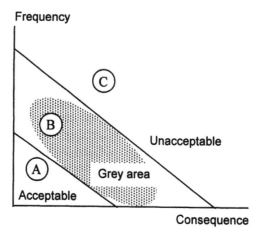

Figure 4.1   Frequency and consequence diagram for evaluation of risks (arbitrary logarithmic scales).

Hazard B is in the grey zone between limits. Should it be accepted or not? This is often a complicated question, especially in large and complex systems. Two general principles are often valid here:

• ALARA As Low As Reasonably Achievable
• ALARP As Low As Reasonably Practicable

Applying the ALARP principle means that the best that can be done under prevailing circumstances must be done. For an identified practicable risk reduction measure, the duty holder should implement the measure unless it can be shown that it is not reasonably practicable. This principle is regarded as valid, for example, in the UK nuclear and offshore industries (Schofield, 1998).

ALARA is similar, but usually interpreted as less rigorous. The risk is reduced as far as reasonable, rather than as far as possible. One interpretation of the ALARA technique is that the costs of safety equipment are balanced against the "values" of the increased safety (Taylor *et al.*, 1989).

The two expressions are rather often confused with each other, and there are somewhat different interpretations.

On application areas

Quantitative assessments are essential, especially when consequences are high and accidents can occur with several people killed. There is a large literature in this area, and there are also international standards (e.g. IEC, 1995). In Europe, the regulation for chemical plants with a potential for major accidents (CEC,

1996) has generated large interest in quantitative assessments and acceptance criteria. Also in the nuclear industry, aviation, and so on such approaches are of high interest.

The approach is also applied to other less dangerous systems and normal accidents, but then there is much wider choice of alternatives (discussed in sections 4.3 and 4.4).

Quantitative risk estimation

Quantitative risk estimation can be done in a number of ways and includes several parts (IEC, 1995). *Frequency analysis* gives an estimate of the likelihood of each identified undesired event. Three general approaches can be used separately or in combination:

- Use relevant historical data.
- Apply analytical techniques, e.g. Fault Tree or Event Tree.
- Use expert judgement.

Consequence analysis estimates likely impact if the undesired event occurs. In the chemical industry, for example, there are a large number of calculation methods for gas emissions, and also for events related to fires and explosions. A detailed account of this type of estimation is beyond the scope of this book, and the reader is referred to the more specialised literature (e.g. Lees, 1996).

The *risk calculations* should help to express the risk in suitable terms. Some commonly used measures are:

- Predicted frequency of mortality to an individual (individual risk).
- Plot of frequency versus consequence for societal risk. Known as the F–N curve, where F stands for frequency and N for the cumulative number of undesired outcomes (e.g. people dying).
- Statistically expected loss rate, in terms of casualties, economic costs, or environmental damage.

There are many uncertainties associated with the estimation of risk. "*Uncertainty analysis* involves the determination of the variation or imprecision in the model results, resulting from the collective variation or imprecision in the parameters and assumptions used to define the model" (IEC, 1995). *Sensitivity analysis* is closely related, and involves the change in response of a model to changes in individual model parameters.

Risk classification

The reasoning above is usually based on computed values of consequences and frequency. In practice, such values are seldom available. Numerical estimates are difficult to make and require considerable effort. This applies particularly to common workplaces where the purpose of a safety analysis is to assess "normal" accidents.

A common approach is to classify identified hazards according to the consequences of related events and their frequency of occurrence. This might be called "semi-quantitative" assessment. The approach generates estimates rather than definitive results, and is based on the judgements of the people doing the classification. Although the approach is sometimes referred to as "qualitative", it fits best into this "quantitative" chapter from a logical perspective.

Table 4.1   Example of the classification of consequences.

Code	Category
0	Not harmful or trivial
1	Short period of sick leave
2	Long period of sick leave
3	Disablement
4	Fatality
5	Several fatalities, major disaster

Table 4.1 provides an example of a scale for consequences divided into six classes. Table 4.2 gives examples of probability values. These have a wide range—from 1 to around 10^{-6} times per day. The scale is designed to include types of disturbances that may be quite common, but also covers events that occur more rarely, such as accidents and catastrophes. There are several examples on similar scales (e.g. Suokas and Rouhiainen, 1993; Taylor *et al.* 1989).

Table 4.2   Example of the classification of probabilities.

Code	Category	Probability*
0	Very unlikely	1 in a 1000 years
1	Unlikely	1 in a 100 years
2	Rather unlikely	1 in 10 years
3	Rather likely	Once a year
4	Likely	Once a month

* Lower limit, i.e. less likely than the specified probability.

Such estimates can be combined into a single measure of risk. Both the classifications given above are logarithmic, and a summary measure of risk is obtained simply by adding the two values.

Using estimated values in risk assessment has the advantage of permitting comparisons to be made between different identified hazards. Their primary advantage is that it forces hazards to be discussed in a fairly systematic manner.

But the approach also has a number of disadvantages. The major one is concerned with probability estimation. It is especially difficult to make a probability estimate for events that only rarely occur. As a guide, it can be said that about one in every 20 industrial workers sustains an occupational injury each year; i.e. there is a score of "2" on the scale given in Table 4.2. This represents the sum of all the different risks at work.

Another problem concerns size of consequence. For example, a fall from two metres may result in almost no injury or be deadly in the worst case. In classification of consequences, it is essential to decide whether a category concerns the "worst case" or some kind of "average value". These problems are further discussed in Chapter 14.

4.3 QUALITATIVE ASSESSMENTS

General

Qualitative assessment approaches are more prevalent in common workplaces, but usually receive less coverage in handbooks about safety. The basic question is whether a workplace should be approved or not. This requires some kind of comparative criteria.

At a general level, laws, regulations, and standards provide a set of criteria that should be met. This is a complicated matter, and is hard fully to interpret in general practice. However, a general risk-assessment strategy might be to demonstrate that compulsory requirements are followed.

Note that the "semi-quantitative" approach to risk classification, which is discussed in Section 4.2 (tables 4.1 and 4.2) is sometimes regarded as a form of qualitative assessment.

Qualitative criteria for risk acceptance

Taylor *et al.* (1989) offered examples of qualitative criteria for risk acceptance. Their list was originally designed for major hazard plants but also has a general application. It encompasses the following:

- A performance requirement, such as strength of components and safety components, e.g. safety valves.

- A fail-safe requirement, implying that certain component failures result in a "safe state".
- A coverage requirement, describing the disturbances for which the safety system should be designed.
- Single and double failure criteria can describe how many different safety systems there should be, in order to prevent specific accidents.
- "Defence in depth", an extension of the single failure criteria originally applied in the nuclear industry area.

These types of criteria imply a special analysis that should demonstrate that the safety criteria are met. Examples of methods related to barriers and safety functions are given in Chapter 10.

Machinery safety criteria

An example of conditions for approving a machine is given in the European standard for risk assessment of machinery (CEN, 1996). In brief, it states that risk reduction can be concluded if the following conditions are met (abbreviated text):

- The hazard eliminated or the risk reduced by design or safeguarding.
- The type of safeguarding selected is appropriate for the application.
- The information on intended use of the machinery is sufficiently clear.
- The operating procedures for use of the machinery are consistent with the ability of personnel who use the machinery.
- The recommended safe working practices for the use of the machinery are adequately described.
- The user is sufficiently informed about the residual risks in the different phases of the life of the machinery.
- If personal protective equipment is recommended, it is adequately described.
- Additional precautions are sufficient.

4.4 DIRECT RISK ASSESSMENT

Requirements for practical risk assessment

Many of the methods described in this book generate lists of hazards. Usually, there are quite a few, and all need to be evaluated. This means that it is advantageous if an evaluation is not too time-consuming.

A risk evaluation should offer advice on each of the identified hazards. It is also intended to give a comprehensive judgement whether the entire system (plant, workplace or machine) is safe enough. Documentation should give information to persons in charge so as to support their later decisions.

Estimations of frequency and severity can be helpful, but the author's experience is that the majority of hazards lie in the intermediate zone between clear acceptance and obvious danger. This means that the ALARA principle, or something similar, needs to be applied.

Laws and regulations are essential to consider in any evaluation, and it is not possible to adopt a pure quantitative approach. At least in large companies, a set of in-house standards is sometimes available. They might state, for example, that safety has a high priority, and that all reasonable safety measures should be taken.

Some methods of safety analysis often generate information that, apart from accident hazards, also concerns health hazards, environment problems, and production disturbances. One acronym is often employed in this context: SHE (Safety, Health and Environment). By adding production we achieve SHEP (Safety, Health, Environment, and Production). If an analysis can handle matters other than safety in a consistent way, it is of clear advantage.

Direct risk acceptance scale

A practical approach is to directly classify each identified hazard into two main categories: those that are "acceptable" and those that are "not acceptable". In concrete terms, "not acceptable" means that a safety measure is required. Table 4.3 shows a risk scale based on that principle.

Table 4.3   Example of a direct risk acceptance scale.

Code	Description
0	Negligible risk
1	Acceptable risk, no safety measure required
2	Safety measure recommended
3	Safety measure essential

Types of consequences

Often in a safety analysis, accidents causing injuries to people are considered. However, an unwanted event may also cause damage to the environment or halt production. In the analysis of a system, conditions can be found that also negatively affect the people working within it. There are several advantages to encompassing safety, health, environmental and production aspects within one and the same analysis. Table 4.4 proposes a classification for such an integrated approach.

Table 4.4   Classification of types of consequences (SHEP).

Code	Description	
S	Safety	Accident hazard for people
H	Health	Health problem for people
E	Environment	Environmental problems
P	Production	Problems with production, quality, etc.

If both these classifications are used, a particular hazard might be classified as "S1, P3". This means that there is an accident risk to people which is classified as acceptable. There is also a production risk, which is important to handle. The combined evaluation then calls for an improvement. An example how this can be applied is shown in Table 7.6.

Criteria for action or acceptance
There is a general need for assessment criteria. Sometimes, they are available as clear directives, but judgement is often involved. Examples of factors to consider are:

- Directives issued by the authorities (breach calls for an improvement).
- Company policy and regulations, which might be more concrete than general directives.
- Compilation of good praxis at similar installations.
- A poor accident record at an installation, which provides grounds for reducing hazards wherever possible.
- Severe consequences, e.g. a person might be killed
- Low system tolerance for *human errors*, e.g. that a single mistake can trigger a hazardous event.
- Low system tolerance for *technical faults*, e.g. that a single failure can trigger a hazardous event.
- That a single failure, or two failures, can trigger an accident (in this case Table 9.2 may be of help in setting up criteria for acceptance).
- Availability of a suitable solution.
- Uncertain available knowledge concerning critical facts or how the system works in reality.
- Cost-benefit considerations.

Directives of authorities sometimes provide a reliable basis for assessment. On occasions, a directive is sufficiently precise for it to be established directly that something must be done about a particular hazard. In such cases, it is appropriate to make a note on the analysis record sheet that refers directly to the section of the directive in question. This should also be done where the company's own regulations apply.

Knowledge can be inadequate, e.g. about how a system works, or what the consequences of an error might be. One approach in this context is to take a break in the assessment procedure and check. Another way (often the best) is to propose a measure ("2" or "3" on the scale), which consists of pursuing a deeper investigation or even a complementary safety analysis. This can also accomplished by adding a further code to Table 4.3. It could be "C", which might refer to "Complementary Investigation or Information Needed".

4.5 PRACTICAL ASPECTS OF RISK ASSESSMENT

Aims of assessment

Risk assessment is an important part of safety analysis. In the planning of an analysis, it is also essential to determine assessment requirements and approaches. These, in turn, are determined by the overall aim of the analysis, and type of object being analysed.

Some examples of aims are given at the beginning of Section 4.1. In practical application they may be either specifically or generally worded.

In the setting of aims it should be noted that assessing accident risks is not a simple task. It cannot be assumed that all analysts will come to the same conclusion. Objective results, those that are independent of the assessor, are impossible to obtain. There is a subjective element to risk assessment, which stems from differences in attitudes and values. Related problems and aspects are further discussed in Chapter 14.

Cost-benefit considerations

It is impossible to get away from financial matters, i.e. what safety measures might cost. Cost-benefit considerations are difficult to make, and it is easy just to see the costs. But including production problems as a parameter in an assessment might improve the situation (see Section 13.6).

Nevertheless, in making a risk assessment, an attempt should be made not to place excessive weight on cost at an early stage. An appropriate time to take up financial issues is when discussions take place on which measures should be implemented, how available resources should be utilised, and which priorities should be set.

Team evaluation

There are advantages to making risk evaluations in a suitably composed team. In the workplace, a team might consist of a supervisor and safety representative, together with a safety professional. Its members then make the

judgement together. Criteria for evaluation ought to be discussed and clarified
before embarking on the risk assessment itself.

The author's own experience of applying "direct risk assessment" in teams
is that it is a quick procedure. It usually takes less than a minute to reach
agreement for a particular hazard, once people get used to the reasoning
involved. Experiences of practical use show that there are relatively few
divergent assessments. In 90–99% of assessments there is direct agreement
within the team.

The differences in judgement are in fact less than might be expected in the
light of the different values that people hold and the various positions they
occupy within a company. However, agreement is not absolutely necessary,
and if it is not reached, this can simply be noted on the record sheet.

Evaluation is used to recommend a final decision to be made by the
company management. It also acts as support for whether proposals for
improvements should be developed or not.

Alternatives

The main alternatives for making risk assessments are:

- Following the relevant specification in a regulation or the like.
- Quantitative assessment.
- Semi-quantitative assessment.
- Direct assessment.
- Assessment of barriers and safety functions (see Chapter 10).

Conclusions

For assessment of risks in a common workplace there appear to be two main
alternatives in most situations. Normally, a choice is to be made between them
if a traditional safety analysis method is to be employed.

Semi-quantitative assessment is perhaps the most common choice—
possibly because it resembles the probabilistic approach, although it has less
accuracy. Also in this kind of assessment, an evaluation needs to be made of
whether a risk is acceptable or not.

Direct assessment is simpler, and it makes clear that assessment is a
judgement and not completely objective. One of its advantages is that it can
fairly easily encompass safety, health, environment and production aspects (see
Table 4.4). The author has arrived at the conclusion that this approach is
favourable in most applications.

The advantages and disadvantages of these two approaches have been
discussed above. Further potential difficulties are discussed in Chapter 14.

5
Energy Analysis

5.1 PRINCIPLES

The idea underlying Energy Analysis is a simple one. For an injury to occur, a person must be exposed to an injurious influence—a form of "energy". This may be a moving machine part, electrical voltage, etc.

In using this method, the concept of energy is treated in a wide sense. Energy is something that can damage a person physically or chemically in connection with a particular event. An injury occurs when a person's body is exposed to an energy that exceeds the threshold of the body. The purpose of the method is to obtain an overview of all the harmful energies in an installation.

The approach of seeing energy as cause of injury was first developed by Gibson (1961) and Haddon (1963). The concept has proved useful, and has been further developed and discussed in several books and reports (e.g. Hammer, 1972; Haddon, 1980; Johnson, 1980). An additional feature of the account presented here is the way in which the analytical procedure is broken down into a number of defined steps (Harms-Ringdahl, 1982).

Thinking in energy terms is based on a model of systems (and of reality) that contains three main components:

1. That which might be harmed, usually a person but it could be equipment or industrial plant.
2. Energies, which can cause harm.
3. Barriers, which prevent harm from being caused, such as safeguards for machinery.

In the model that applies to accidents, a person or object comes into contact with a harmful energy. This means that the barriers have not provided sufficient protection.

Harmful energy can take on many forms, such as an object at a height (from which it may fall) or electrical voltage, i.e. energies in a traditional sense. By adding acutely poisonous and corrosive substances, etc. a fairly comprehensive picture of the injuries that might affect a human being is obtained. Table 5.2 provides a summary of different kinds of energies.

One essential part of the energy model is the concept of barriers. These will prevent the energy from coming into contact with the person and/or cause injury. Table 5.3 shows various safety measures that might prevent accidents from occurring due to the release of energies. These measures can also be seen as barriers, which might already exist or could be introduced.

5.2 ENERGY ANALYSIS PROCEDURE

An Energy Analysis contains four main stages, as well as making preparations and concluding the analysis. It is usually best fully to complete each stage before moving on to the next. As an aid to analysis, a specially designed record sheet can be used (example in Table 5.4).

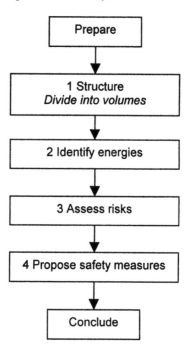

Figure 5.1   Main stages of procedure in Energy Analysis.

PREPARE
Before embarking on the analysis itself, a certain amount of preparation is required. This concerns a definition of the limits of the "object" of study, which may be a single machine, a workplace, or a whole factory. During preparation other clarifications may also be needed, e.g. with regard to what assumptions should be made concerning the machine. This is similar to other methods of safety analysis, and is discussed in Chapter 13.

One essential aspect is to obtain information about the installation being considered. For Energy Analysis, this can consist of technical drawings and photographs. If the installation already exists, you can simply go round and look at it.

1. STRUCTURE
The purpose of the structuring stage of the analysis is to divide the system into suitable parts, which are then analysed one at a time.

In general, structuring is performed in accordance with the physical layout of the installation under study. In principle, the plant or equipment is divided into "volumes" (spatial segments). If the analysis is applied to a production line, it is appropriate to go from one end of the line to the other. The installation can be envisaged as being divided up into "boxes".

This means that the boundaries of the entire system to be analysed should also be thought of in volume terms. After structuring, a check should be made as to whether any component has been omitted or "lost" in some way. If the entire area to be analysed is not covered, supplementary volumes are needed. Sometimes, it may be wise to add an extra "volume" to cover anything lying outside the area where the object in question is located.

2. IDENTIFY ENERGIES
For each "box" or volume, sources and stores of energy are identified. The checklist of energies shown in Table 5.2 can be used as an aid to this.

One problem is to determine the lowest level of energy with which the analysis should be concerned. There is a trade-off between comprehensive coverage of hazards and the avoidance of trivia. This decision should be made in the light of the aims and level of ambition of the application in question. However, an energy should not be excluded just because it seems unlikely that a human being will be exposed to it.

3. ASSESS RISKS
Each identified source of energy is assessed. The assessment can be made in different ways, as discussed in Chapter 4. The method itself does not prescribe what kind of assessment should be made. With energies, it is quite natural to think in terms of sizes of consequences in terms of injuries, and the classification in Table 4.1 can be used. It might be applied on its own, or some kind of probability assessment could be added. One difficulty is that any one energy may have a variety of consequences.

In the assessment, identifying the presence and efficiency of barriers is essential. They will affect the seriousness and likelihood of injuries. One way is to make a direct risk assessment (see Section 4.4), which could then employ the scale in Table 4.3. For application in Energy Analysis, Table 5.1 offers some comments to aid interpretation. In principle the choice is to accept the system as it is, or determine that safety improvements are needed.

Table 5.1   *Example of a direct risk acceptance scale applied in Energy Analysis.*

Code	Description	Comment
0	Negligible risk	Energy cannot cause any significant injury.
1	Acceptable risk, no safety measure required	Energy can cause injury, but barriers are adequate.
2	Safety measure recommended	Barriers should be improved.
3	Safety measure essential	Serious consequences and inadequate barriers.

4. PROPOSE SAFETY MEASURES

At the next stage, a study is made of the energies for which safety measures are required (coded 2 or 3 as in Table 5.1). Questions are raised concerning whether and how risks can be reduced. Can a particular energy be removed or reduced? Can safety devices be installed? Table 5.3 shows a methodology that can help in finding safety measures. The example in Section 5.3 illustrates how the checklist can be used.

At the beginning, it is a matter of generating and sifting through ideas. It is good to be able to suggest a variety of solutions, since it is not certain that the first will be the most effective. Then, the most suitable solutions can be selected.

CONCLUDE

The analysis is concluded by preparing a report, which summarises the analysis and its results. It might contain descriptions of the limits and assumptions of the analysis, the most important energies, and proposals for safety measures. Sometimes, a record sheet might suffice.

Energy checklist

Table 5.2 shows a checklist of different types of energies. It is designed for use as an aid to identification. For most categories, the link between energy and injury is obvious. But some types of energies may require further comment.

Table 5.2   Checklist for Energy Analysis.

1. POTENTIAL ENERGY	6. HEAT & COLD
Person at a height	Hot or cold object
Object at a height	Liquid or molten substance
Collapsing structure	Steam or gas
Handling, lifting	Chemical reaction
	Condensed gas (cooled)
2. KINETIC ENERGY	**7. FIRE & EXPLOSION**
Moving machine part	Flammable substance
Flying object, spray, etc.	Explosive:
Handled material	– material
Vehicle	– steam or gas
	– dust
	Chemical reaction, e.g.:
	– exothermic combinations
	– impurities
3. ROTATIONAL MOVEMENT	**8. CHEMICAL INFLUENCE**
Machine part	Poisonous
Power transmission	Corrosive
Roller/cylinder	Asphyxiating
	Contagious
4. STORED PRESSURE	**9. RADIATION**
Gas	Acoustic
Steam	Electromagnetic
Liquid	Light, incl. infra and ultra
Pressure differences	Ionised
Coiled spring	
Material under tension	
5. ELECTRIC	**10. MISCELLANEOUS**
Voltage	Human movement
Condenser	Static load on an operator
Battery	Sharp edge
Current (inductive storage and heating)	Danger point, e.g. between rotating rollers
Magnetic field	Enclosed space

"Chemical influence" is treated here as an "energy" that might give rise to injury. In some cases, it is possible to conceive of this influence in terms of the chemical having a micro-level effect on human cells. "Asphyxiating" chemicals are gases or liquids that are not poisonous in themselves, but which restrict or eliminate access to air. This subcategory might refer to the possibility of being exposed to a suffocating gas or of drowning in water.

Wider energy perspective

The final category on the checklist is headed "Miscellaneous". It is included to provide an additional check on whether the focus of the analysis has been too narrow in a technical sense. An extra check is obtained on whether the movements of a human being might involve the risk of falling, stumbling, colliding with protruding objects, etc. "Static load" may help to identify work situations where a person is operating in a poor ergonomic position.

The "Sharp edge" and "Danger point" subcategories do not really refer to forms of energy, but such items can be seen in terms of energy concentrations when a person or piece of equipment is in motion. "Enclosed space" can be used as an extra check. There might be overpressure, toxic gases, etc., which are not normally dangerous but could be under unusual circumstances.

The list contains a few deliberate inconsistencies. Some categories do not refer to energies in a physical sense, but are conceived as such as a means for getting at sources of risk (hazards). For example, "Collapsing structure" may apply where the object under study is a heavy installation (such as a liquor tank). The energy in question is the potential energy of the tank. The category "Collapsing structure" draws attention to major items of equipment and the possibility that they overturn or have other defects.

Similarly, the subcategory "Handling, lifting, etc." is used to cover the potential and kinetic energy of a handled object. The idea is that problems related to the manual handling of materials can also be treated at the identification stage.

A systematic approach to safety measures

One of the advantages of Energy Analysis is that it provides a systematic means for developing safety measures (Haddon, 1980; Johnson, 1980). Table 5.3 provides examples of measures designed to reduce risks that are generated by adopting the energy approach.

Table 5.3   Finding safety measures using Energy Analysis.

Safety measure	Examples
The energy 1. Eliminate the energy	Work on the ground, instead of at a height. Lower the conveyor belt to ground level. Remove hazardous chemicals.
2. Restrict the magnitude of the energy	Lighter objects to be handled. Smaller containers for substances. Reduce speed.
3. Safer alternative solution	Less dangerous chemicals. Handling equipment for lifting. Equipment requiring less maintenance.
4. Prevent the build-up of an extreme magnitude of energy	Control equipment. Facilities for monitoring limit positions. Pressure relief valve.
5. Prevent the release of energy	Container of sufficient strength. Safety railings on elevated platforms.
6. Controlled reduction of energy	Safety valve. Bleed-off. Brake on rotating cylinders.
Separation 7. Separate the object from the energy flow: a) in space	One-way traffic. Separate off pedestrians and traffic. Partition off dangerous areas.
b) in time	Schedule hazardous activities outside regular working hours.
8. Safety protection on the energy source	Machine safeguards. Electrical insulation. Heat insulation.
Protection of the object 9. Personal protective equipment	Protective shoes, helmets.
10. Limit the consequences when an accident occurs	Facilities for stopping the energy flow. Emergency stop. Emergency shower facilities. Specialised equipment for freeing a person (if stuck).

5.3 EXAMPLE

In this example, a tank for the storage of sodium hydroxide (lye) is to be acquired. There is a desire to make a preliminary assessment of the hazards involved. A starting point is a sketch of the installation, which provides the basis for a simple Energy Analysis.

System description

Concentrated sodium hydroxide is to be stored in a stainless-steel tank. At lower temperatures the lye is viscous, and heating equipment using an electrical current is needed. The tank is filled using a tube equipped with a valve. On top of the tank there is a manhole and a breather pipe. Under the tank there is a pit. A ladder has been permanently installed to provide access to the tank. Not visible on the sketch is a tube with a tap, used to tap off the liquid.

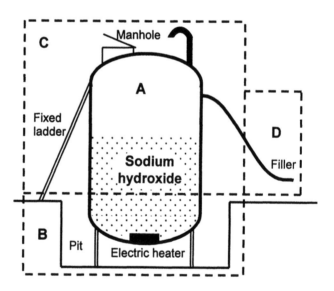

Figure 5.2   Liquor tank.

Preparing

Our start material consists of the sketch and the description above. The limits of the system are set by what is visible on the sketch. Table 5.4 shows an example of the record sheet used.

Analysis

1. STRUCTURE
A classification is made into four volumes, as shown in Figure 5.2:
A.   The tank.
B.   The pit (the space under the tank).
C.   The area surrounding the tank.
D.   The filler tube and its surroundings.

2. IDENTIFY ENERGIES
Let us start with the tank (Volume A) and follow the checklist (Table 5.2).
First, there is *Potential energy* (1).

- "Person at a height" will be relevant when someone goes down into the tank for servicing.
- The level of the liquid is above that of the tapping-off tube. If the valve is opened or if a connecting tube fails, the lye will run out.
- The tank has great mass. It requires stable supports ("Collapsing structure").

Then, *Stored pressure* (4) may be relevant. If the ventilation system fails, there will be high pressure when lye is pumped in, and the whole tank might burst. When tapping-off lye there may be low pressure, but this is not harmful. (The pressure of liquid was earlier treated as a form of potential energy.)

Electric (5) refers to the electric-power supply to the heating element. Insulation failure is hazardous. Lye is electrically conductive.

Heat & cold (6) applies to the heating element. Over-heating might occur if the liquid level is low, or electric power is not turned off correctly

Chemical influence (8) is obviously relevant because lye is highly corrosive. It is certainly the most obvious and the greatest hazard in the system.

The *Miscellaneous* (10) category provides an opportunity for a variety of items to be taken up. It is not necessary to think strictly along energy lines. For example, one might wonder about the manhole. It has to be large enough, and there must also be space for a ladder.

Then we continue with the pit (Volume B) and the remaining volumes.

3. ASSESS RISKS
The risks are assessed, and the scale in Table 5.1 is used. The assessments are shown on the record sheet (Table 5.4). The judgement in this example reflects the thoughts of the imaginary study team.

4. PROPOSE SAFETY MEASURES
Before making a start on conceiving safety measures, the possibility of grouping the hazards into more general categories should be investigated. Lye

comes up throughout the analysis, and all cases where it occurs might be considered simultaneously. We imagine that a number of technical safety measures will be conceived and proposals for job routines made. These will apply to lye in general. In the table, the term "lye package" is used to refer to this set of safety measures. In addition, a check should be made on each place where lye is noted on the record sheet to see if extra control measures should be taken.

It can be difficult to make concrete safety proposals for some hazards. In such cases, it might then be stated, for example, that job routines must be established or that a further investigation needs to be made. Or, it is even possible to note that no solution has been thought of, but that the problem still requires attention.

Let us take the hazards created by the lye itself as an example of how a systematic approach to the consideration of safety measures can be employed.

1.	Eliminate	Can lye be removed from the process?
2.	Restrict	A smaller tank?
3.	Safer alternative solution	Can the lye be replaced by another chemical, or can a diluted mixture be used?
4.	Prevent build-up	Safety device to prevent over-filling?
5.	Prevent release	Secure connection for hose on filling, and a method for emptying the filler tube after the tank has been topped up.
6.	Control reduction	Overflow facilities in case of over-filling.
7.	Separate off the human being	Prohibit unauthorised entrance and fence off the area.
8.	Safety protection on the object	Keep the filler tubes in a locked cupboard.
9.	Personal protective equipment	Protective clothing.
10.	Limit the consequences	Emergency shower facilities. Water for flushing. Draining facilities. Emergency alarm. First-aid facilities. Make the pit under the tank sufficiently large.

Table 5.4 Record sheet from the Energy Analysis of a tank with lye.

Volume / Part	Energy	Hazard / Comments	Evalu-ation	Proposed measures
A. Liquor tank	Person at a height (4 metres)	Falls down / *During service*	2	Routines for service
	Level of lye	Contact with lye / *Lye can run out*	3	"Lye package"
	Weight of tank (10 tons)	Falls or collapses / *If damaged or poor design*	1	(Standard construction)
	Excessive pressure	Tank rupture / *On filling*	0	(Good ventilation)
/Heating system	Electric (380 volts)	Shock / *If poor insulation, lye is conductive*	2	Check proposed installation
	Heat	Burn injury	1	
/General	Lye (10 tons)	Injury to eyes and skin, corrosive	3	"Lye package"
B. Pit	Height (1.4 metres)	Falls	3	Railings and fixed ladder
/Heating system	See above	–	–	
C. Outside tank	Person on platform or ladder	Falls	3	Suitable design of platform, railings and fixed ladder
	Tools and equipment at a height	Fall on people below	2	As above
D. Filler tube	Level of lye	Contact with lye remaining in tube	3	"Lye package", routines, and lock

Concluding

After completing the analysis, a summary is prepared. In this case, it will consist of a list with recommendations to be applied during continuing design and planning.

Remarks

The results of the analysis are not remarkable, but they provide a more complete picture than otherwise would have been available. In this example, it would have been possible to stare blindly just at the hazards created by the lye itself. Then, probably, only some of the problems would have come to light, and only some conceivable safety measures suggested. A more extensive analysis would have dealt with situations that arise in the course of re-filling the tank, etc. Therefore, in this case, a supplementary method should be employed.

5.4 COMMENTS

A simple method

The method can seem cumbersome, involving long checklists for each stage of the analysis. But, with a little experience, it is simple and quick to use. The identification stage can be completed in just one or a few hours, even with quite large systems.

Examples of pitfalls when using Energy Analysis:
- Some "volumes" are missed.
- Too much time is spent on details, e.g. trivial energies.

Identifying energies

Sometimes, judgement is needed in weighing up which energies should be included in the analysis. In principle, anything that can lead to an injury to a human being should be included. That contact with an energy is unlikely is no reason for it not to be included

On the record analysis sheet, a note should be made of which specific energy is concerned. Simply repeating the names of categories on the checklist should be avoided. For example, "lye" and not just "corrosive substance" should be written, and "tool at a height" should be entered rather than "object at a height". It is also good to note the magnitude of the energy, e.g. how many metres, or weight in tons. This will help in the assessment of risks.

Assessing risk

The most practical way of proceeding is to identify energies in all volumes before embarking on the next stage of the analysis. This permits a better overall picture to be obtained and a more consistent form of risk assessment to be applied.

In the example of the liquor tank, the analysis was based on the idea of assessing whether the risk was acceptable or not. In Energy Analysis, a good alternative is to assess hazards using a scale of measures of conceivable injury, e.g. whether the magnitude of the energy would lead to fatal, serious, minor or trivial injury (for further discussion, see Chapter 4 and Table 5.1).

Ideas for safety measures

In order to generate ideas for safety measures, it is best to think freely and try to come up with as much as possible. The checklist is designed to be an aid to the imagination and a means for getting away from rigid lines of thinking. It is meant to provide different angles of approach. When a body of ideas has been built up, then the process of sifting and improving the ideas can begin.

Energy magnitude

In many cases it is possible to specify the magnitude of the energy with which one is concerned, e.g. in terms of its height, weight, or speed. This provides a more concrete basis for the assessment of risk.

With respect to rotating objects, this can be difficult. Then, however, it is possible to convert rotational energy into one of its equivalent forms. What would it mean in terms of velocity if the object came loose? To what height could the object be lifted? In principle, the moment of inertia could be calculated, and, on the basis of this and the rotational energy, the magnitude of the energy could be derived.

In most cases, however, a simple calculation, which has the advantage of being easy to remember, can be made. If the rotating object has the bulk of its mass on its periphery, e.g. a tube that rotates along its longitudinal axis, the magnitude of the energy is obtained from the familiar expression:

$$W = m \, v^2/2 \qquad (5.1)$$

Where W stands for energy (in joules), m is the mass (kg), and v is the circumferential velocity in metres per second. The equivalent height h is given in Equation 5.2, where g stands for acceleration due to gravity, usually 9.8 m/s^2.

$$h = v^2/2g \qquad (5.2)$$

In the case of a solid cylindrical object, the moment of inertia is half that of a hollow cylinder. The following expressions are then applicable:

$$W = m\ v^2/4, \text{ and} \tag{5.3}$$

$$h = v^2/4g \tag{5.4}$$

Let us take the example of a paper-rolling machine. Paper is wound onto a reel, and a reel can weigh up to several tens of tons. Paper may be rolled at a speed of 2000 metres/minute. The equivalent potential energy is that of mass at a height of 28 metres.

6
Job Safety Analysis

6.1 PRINCIPLES

In Job Safety Analysis, attention is concentrated on the job tasks performed by a person or group. The method is most appropriate where tasks are fairly well defined. The analysis is based on a list of the phases into which a job task can be broken down. The approach consists of going through the list point by point and attempting to identify different hazards at each phase. Sometimes, this procedure is called Job Safety Analysis, sometimes Work Safety Analysis.

The method is not based on any explicit model of how accidents occur. The production system is seen from the perspective of either the worker or the job supervisor. It is divided up into tasks controlled by machines and those governed by job instructions. However, the picture of accidents is fairly close to the energy model, and some descriptions of the method contain checklists of different energies.

One of the advantages of the method is that it is straight-forward and relatively easy to apply. Several descriptions of the method have been published (e.g. Grimaldi, 1947; McElroy, 1974; Heinrich *et al.*, 1980; Suokas and Rouhiainen, 1984).

6.2 JOB SAFETY ANALYSIS PROCEDURE

Figure 6.1 shows that Job Safety Analysis consists of four main stages, plus a preparatory and concluding part. It is recommended that each is completed in sequence.

PREPARE
Preparation includes defining and setting the boundaries of the job tasks to be analysed, and gathering information on instructions where these are especially important for the implementation of a task. For the analysis, a special record sheet is used. An example is shown in Table 6.1.

In this type of analysis it is beneficial to involve a team of people in the workplace. The team might include someone familiar with the method, a job supervisor, and a person who knows the job in practice and its potential problems.

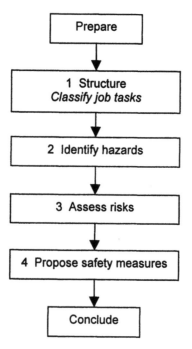

Figure 6.1   Main stages of procedure in Job Safety Analysis.

The main reasons for engaging a team are:

- Getting better information about the job and its conditions.
- Obtaining a broader perspective on risk assessment and proposals for measures.
- Improving circulation of results.
- Having better confidence in the results obtained.

1. STRUCTURE

The purpose of the structuring stage of the analysis is to obtain a list of work tasks. A suitably detailed list of the different phases of the work under study is prepared. Good basic material consists of standard job instructions, but these should be regarded as a starting-point. Usually, they cannot be assumed to be either complete or correct. It is important to take account of exceptional tasks and those that are only seldom undertaken. The following items should be considered:

- The standard job procedure.
- Preparations for and finishing off the work.
- Peripheral and occasional activities, such as obtaining materials, cleaning, etc.
- Correcting the disturbances to production that might arise.
- The job as a whole, including relations to descriptions, planning and other related tasks.

Depending on type of work, the following two components may also be included:

- Maintenance and inspection.
- The most important types of repairs.

2. IDENTIFY HAZARDS

The subtasks on the list are gone through one at a time. A number of questions are posed in relation to each of these:

- What types of injuries can occur?
    - Pinch/squeeze injuries or blows, moving machine parts, objects in motion or at a height, etc.
    - Cuts or pricks/stabs, sharp objects, etc.
    - Falls, working at a height, etc.
    - Burns
    - Poisoning
- Can special problems or deviations arise in the course of the work?
- Is the job task difficult or uncomfortable?
- Is the task usually done in a different way then prescribed, or are there incentives to deviate from regular procedures?

It is advantageous not to restrict the analysis to accidents alone. Contact with chemicals, ergonomic problems, etc. may also be included, which may increase the benefits of the analysis.

3. ASSESS RISKS

Each identified hazard or problem is assessed. A variety of approaches to classification and risk assessment may be utilised (for further discussion, see Chapter 4). The method itself does not prescribe what kind of assessment should be made.

4. PROPOSE SAFETY MEASURES

The next stage of the analysis is based on the hazards regarded as serious. When going through the record sheet, an attempt is made to propose ways of reducing risks. Such measures may apply to:

- Equipment and task aids.
- Work routines and methods. (Can the work be carried out in a different way?)
- Elimination of the need for a certain job task.
- Improvements to job instructions, training, etc.
- Planning how to handle difficult situations.
- Safeguards on equipment.
- Personal protective equipment.

This stage of the analysis principally concerns the creation of ideas. It is of benefit if ideas for several alternative solutions are generated. Several measures may be required to reduce a given risk. A particular safety measure may be hard to implement, so an alternative might be needed.

Several different items that are similar in one way or another may be merged into one, e.g. if the hazards have similar causes or if a common safety measure is required. The proposed measures are entered on the record sheet.

CONCLUDE

The analysis is concluded with a summary of results. In simple cases, the record sheet itself may be used to report the results. The list of job tasks and the record of the analysis may also be used, fairly directly, to produce an improved set of job instructions.

6.3 EXAMPLE

Job Safety Analysis is featured in some of the examples presented later in this book (Chapter 15). The example shown here concerns an analysis of a machine for the rolling and cutting of paper (from Section 15.2). In principle, standard production at the machine consists of three simultaneous operations. First, there is the unrolling of a wide reel of paper; then, the paper passes through a set of rotating knives; finally, a set of narrower reels are wound. Figure 6.2 shows work at such a machine, with the wide starting reel to the right.

Jobs in the workplace encompass the following principal tasks:

1. Removal of produced reels and transportation to store room.
2. Preparation of machine for new production cycle.
3. Installation of new base reel in machine.
4. Operation of machine.
5. Tasks at start and end of workday.
6. Corrections and cleaning.
7. Other transport of materials.

The first four tasks (1–4) are related to regular planned work. The three final ones are less well planned, and may vary quite considerably by nature. During the analysis all important deviations that need correction should be directly included in the specific task to which they belong.

Table 6.1 shows an extract from a record sheet. It concerns the task "*2 Prepare machine for new production cycle*". The task has been subdivided into four parts. Figure 6.2 illustrates the part of job task 2.4 when the operator is feeding new paper through the machine. It can be advantageous to break down the tasks in even larger detail at difficult or dangerous situations. In the example this is indicated by a slash "/Correction of disturbances".

Figure 6.2   Work at paper-rolling machine (part of job task 2.4 in Table 6.1).

Record sheets can be designed in several ways. In this example a simple approach is chosen, involving five columns. The column "Comments" is intended for explanations so that the reader can understand what might happen, why it is dangerous, etc. Alternative or extra headings in a record sheet could be *Causes, Consequences*, and *Responsible for measure*.

In the assessment of hazards, a direct approach to evaluation is chosen. It is the same as in tables 4.3 and 5.1. Code 2 stands for "Safety measure recommended" and 3 for "Safety measure essential". In the example, only serious hazards have been entered. Also, a number of proposed measures have been given.

6.4 COMMENTS

Simplicity

The method is easy to learn. But, if the training period is too short, problems may arise as a result of inadequate coverage. Similarly, the range of safety measures proposed may be narrow, compared with that generated when a more highly skilled person is involved.

Simple analyses can be conducted with little preparation and only a small amount of effort. But, if the work to be analysed is more extensive or involves a lot of variation, then there is a need for the application to be more formal, and possibly for the assistance of experienced people.

An advantage of the method is that it is based directly on the ordinary job tasks, which are easy to visualise. It is also based on commonly accepted ideas with regard to safety and regular safety work. For this reason, it is easy to teach the method and get it accepted for direct use by job supervisors and work teams. But that the method is based on a standard approach to safety matters may also be a disadvantage. It makes it harder to avoid having a blinkered view on the work involved.

Applications

The method is useful when applied to more or less manual jobs. These could be machine operating in an industrial workplace, in building work, during repair, etc. It is probably less suitable in highly automated production, when teams have to co-operate, or in general when awareness of the characteristics of the whole system is essential.

One common application of the analytical method is in job planning. A supervisor may consider a repair that has to be made. He goes through his list of what is to be done with the repairers. This enables them to identify hazards at various work phases and determine which safety measures are needed.

Such an analysis is informal, and the records of the analysis, etc. are not so important. A decision can be made immediately, and people appointed to take responsibility. The extra time required for the analysis may be about an hour. Quality of the analysis may not be so high, but the reduction in risks can still be significant.

Information materials

Information is needed to prepare the list of work phases, and to identify hazards. Where systems have been in operation for some time, there is a body of experience available. Generally, this is possessed mainly by those who work directly with the equipment and by job supervisors. This knowledge can be accessed through discussions in a suitably composed study team.

Table 6.1.   Extract from the record of a Job Safety Analysis applied to a paper-rolling machine.

Job task / Part	Hazard	Comments	Evalu-ation	Proposed measures
2 Prepare machine for new production cycle				
2.1 Removal of old base reel	Reel falls down	Rather heavy (40 kg). The reel can get stuck, or the operator can lose grip	2	Lifting equipment and adjustable holding facilities
2.2 Taking away packing band on new reel	Cutting injury	Sharp steel band	2	Use of proper tools
		Packing band recoils, due to tension		Include this task in instruction manual
		Erroneous method (knife)		
2.3 Installation of new base reel	Reel falls down	Heavy (2 tons). Mistake in operating lifting equipment is hazardous	3	Improve instructions
	Squeeze injury	Moving machine parts	3	Improve machine guards
2.4 Feeding new paper through machine	Operator falls, if paper tears	Heavy task, if brakes are not completely off, or if base reel is oval	2	Improve design for releasing brake pressure
	Squeeze injury	Paper reels and steel rollers rotating with great force	3	Develop automatic paper feeder, or change work routines (use the previous sheet of paper to pull through a new reel)
/Correction of disturbances	Squeeze injury	Disturbances often occur	3	Improve automatic control system
	Cuts from roll knife			Develop safer correction methods
				Include in manual and during job training

The information needed can also be obtained from:

1. Interviews.
2. Written job instructions (sometimes incorrect, always incomplete).
3. Machine manuals.
4. Work studies, if these are available.
5. Direct observation, the observer simply standing and watching.
6. Photographs, both to depict problems and to facilitate discussions within the study team.
7. Video recordings, which are especially valuable for tasks that are only seldom undertaken, but have the disadvantage that they take a long time to make.
8. Accident and near-accident reports.

List of phases of work

An important part of the analysis consists in producing a list of job tasks. Sometimes this can take longer than the identification of the hazards themselves. Only brief descriptions of the different phases are needed. It is more important that the list is sufficiently complete. This is checked before the identification stage is embarked upon.

One common dilemma arises when there is a major discrepancy between job instructions and how the job is carried out. This can be a serious problem, which needs careful consideration. The method itself does not solve this problem, but it can be of good help in identifying discrepancies and the hazards these cause.

Time taken by the analysis

The time taken by an analysis may vary considerably, but the method can be regarded as relatively quick to apply. How much time is needed for any one analysis depends on:

- The magnitude/diversity of the task to be analysed.
- The efficiency with which the analysis is conducted and participants are trained.

A rule of thumb is that the identification of hazards takes 5 minutes per phase of work. The number of work phases may come to between 20 and 50. The identification stage of the analysis may therefore be expected to take between one hour and half a day. The author's personal experience is that it takes roughly the same amount of time to produce a list of work phases, and the same time again to conduct discussions on safety measures. In total, the analysis may take between half a day and two days.

7
Deviation Analysis

7.1 PRINCIPLES

Aim

Deviation Analysis is used to study a production system and the activities within it. The aim is to identify and analyse deviations that can cause accidents or other problems. The method includes the development of preventive measures.

Fundamentals

Systems do not always function as planned. There are disturbances to production, equipment breaks down, and people make mistakes. There are deviations from the planned and the normal. Deviations can lead to defective products, machine breakdowns or injuries to people.

The fundamentals of Deviation Analysis can be summarised in a few points. The rationale behind these is explained later in this chapter.

- Accidents are always preceded by deviations; consequently, deviations increase the risk of accidents.
- Knowing the potential deviations in a system enables better understanding of the causes of accidents.
- The risk of accidents can be reduced if deviations are identified and can be eliminated or controlled.
- The production process and the activities it involves make up the object of Deviation Analysis.
- A production system consists of technical, human and organisational elements.
- Deviations are of several kinds; it is essential to consider technical, human and organisational deviations.
- The same principles can be used for the analysis of a system and for the investigation of an accident.

Applications

The method was initially developed specifically for application to accidents. However, the deviation principle has been further developed to turn Deviation Analysis into a more generic method (Harms-Ringdahl, 1982, 1987b). It can be applied to a variety of problems and different types of systems. In any analysis, it is often suitable to handle consequences related to safety, health, environment and production in the same exercise (see also Section 4.4).

The most important application is to analyse production systems in terms of deviations in order to anticipate what might happen. This procedure is called Deviation Analysis and is described in Section 7.3.

However, the principles can also be employed for the investigation of accidents and near-accidents (described in Section 7.5). When applied in this way the analysis is called Deviation Investigation. There may be advantages for a company in using matching methods for safety analysis and accident investigations. In the following section, the discussion focuses mainly on common features of the two applications.

The deviation model of reality

Every method of safety analysis is based on some kind of representation (or model) of the system to be analysed. Essential to Deviation Analysis is the perspective on the system, and it focuses on:

- The combination of technical, human and organisational elements.
- Functions and activities.

This means that modelling of the system is an essential component in the analysis ("structure"). This is a key feature of several methods, e.g. Task Analysis, and is discussed in several places in this book.

Four types of safety measures

Deviation Analysis includes a simple but systematic method for the generation of safety measures. To increase the safety of a system, improvements should be based on deviations that have been identified as hazardous. Efforts are made to generate measures which can:

1.   Eliminate the possibility that a certain deviation will arise.
2.   Reduce the probability that it will arise.
3.   Reduce consequences if it arises.
4.   Support early identification of the deviation and provide for plans on how it should be corrected in a safe and effective manner.

Comment

The concept of deviation is used in a number of other methods, especially in HAZOP (Chapter 8) and Failure Mode and Effects Analysis (Section 11.2). In these there is a technical modelling of the system, and deviations are mainly of physical and technical type.

7.2 ON DEVIATIONS

Definitions

It is fairly easy to understand intuitively what deviations are and why they are important. In the analysis of hazards, a strict definition is not always necessary. Nor may it always be desirable. The purpose of discussing deviations in an analytical context is to discover factors that might lead to hazards. The need for precise definition is greater in some cases, e.g. when a statistical classification is to be made.

A deviation is defined as an event or a state that diverges from the correct, planned or usual function. The function can be a process, a technical function, or a human or organisational activity.

Terminology

Deviations of one type or another appear in several methods for safety analysis. These are discussed in Chapter 12. In the area of accidents, there are a large number of terms used to denote deviations of one type or another. Some examples are disturbance, breakdown, fault, failure, human error and unsafe act.

The deviation concept is a common element in a number of different theories and models. There are a variety of areas of application and definitions (Kjellén, 1984, 2000). On a general level, a system variable is classified as a deviation when its value lies outside a norm. Some system variables are:

- Event or act (i.e. part of a procedure or a human action).
- Condition (i.e. state of a component).
- Interaction between the system and its environment.

Examples of the types of norms that appear in the literature include:

- Legal—a standard, rule or regulation.
- Adequate or acceptable.
- Normal or usual.
- Planned or intended.

Consequences of deviations

Many kinds of deviations can arise within a system. Consequences can be of a large number of different types, some leading to increased risk, others being harmless. Table 7.1 shows a classification of consequences of deviations.

Consequences of Type 1 lead directly to an accident; however, they are usually preceded by several other deviations. Types 2 to 5 increase risk in the system. A "latent failure" can exist within a system for a long time without it being noticed, but when the function is needed it does not work. One example is a defective fire alarm; if a fire starts the alarm is of no help.

A further aspect of the classification concerns the extent to which the deviations can be corrected so that the system can be returned to a safe state. It should be noted that the majority of deviations are neither discovered nor corrected. They do not necessarily have a negative effect.

Table 7.1   Classification of deviations by type of consequences.

Consequence	Comments
1. Direct accident	The deviation leads directly to an accident. For example, the wing of an aeroplane falls off in mid air.
2. Potential accident	An accident occurs if people are in the danger zone or if the system is at a certain operational phase.
3. Indirect accident, as a consequence of the deviation	The deviation in itself does not lead to an accident but could start a chain reaction.
4. Accident in combination (latent failure)	In addition to the deviation itself, certain other independent conditions must be met for an accident to occur.
5. Increase in the probability of an accident	The deviation in itself does not mean that an accident will occur, but it increases the probability of hazardous deviations.
6. Not dangerous	The deviation does not lead to an increase in risk.
7. Increase in safety	The deviation increases safety. For example, someone does not follow an unsuitable job instruction, but works in a safer way instead.

Combinations of deviations

The relation between deviations and increased risk/the occurrence of accidents can be complex. There might be a series of deviations, one leading to another until an accident occurs. This perspective is sometimes referred to in terms of "Domino Theory" (Heinrich, 1931). A more realistic approach is to consider a (large) number of simultaneously existing deviations, which might eventually combine to cause an accident.

The application of Deviation Analysis will usually reveal several potential deviations in a system. These can combine in different ways. An important feature in some safety analysis methods, such as Fault Tree and Event Tree, is the examination and clarification of relations between deviations.

Checklist for Deviation Analysis

A checklist, as shown in Table 7.2, is used for analysis and investigation. It is designed as an aid for the identification of deviations, and is based on technical, human and organisational functions. It should not be seen as a taxonomy of deviations, for it is based on functions of the system. There can be overlaps between categories.

Comments on technical deviations

T1   *General function.* These are deviations from the normal, intended or expected functioning of the system. The system does not work as expected, and there can be different kinds of disturbances. Deviations related to automatic functions and computer control are also included in this category. Examples include a failure to achieve a desired final outcome, that plant stops unexpectedly, or that a machine runs too quickly (see also Table 7.3.).

T2   *Technical function.* These are deviations of a technical nature, e.g. technical failure of a component or module, or interruption of electric power supply.

T3   *Material.* This applies to material that is used in the system, and also to transport and waste. Deviations can concern poor quality, wrong quantity, wrong delivery time, etc.

T4   *Environment.* This refers to abnormal or troublesome conditions in the indoor or outdoor environments. Examples include faulty or dim lighting, bad weather, the accumulation of waste, and other temporary environmental states that give rise to difficulties.

Safety analysis – Principles and practice

Table 7.2   Checklist for system functions and deviations.

Function	Deviation
Technical	
T1. General function	Departure from the normal, intended or expected functioning of the system.
T2. Technical function	Failure of component or module, interruption to energy supply, etc.
T3. Material	Poor quality, wrong quantity, wrong delivery date, etc.
T4. Environment	Waste, poor light, bad weather, any temporary disruptive state of the environment.
T5. Technical safety functions	Machine guards, interlocks, monitors, etc., which are missing, defective or inadequate.
Human	
H1. Operation/movement	Slip or misstep in manual tasks.
H2. Manoeuvring	Lapse or mistake in control of systems.
H3. Job procedure	Mistake, forgetting a step, doing subtasks in the wrong order.
H4. Personal task planning	Choosing an unsuitable solution, violations of rules and safety procedures and risk-taking.
H5. Problem solving	Searching for a solution in a hazardous way.
H6. Communication	Communication error with people or the system, on either sending or receiving a message.
H7. General	Inconsistency in system demands on personnel, concerning skills or knowledge.
Organisational	
O1. Operational planning	Non-existent, incomplete or inappropriate.
O2. Personnel management	Inadequate staffing, lack of skills.
O3. Instruction and information	Inadequate or lacking. No job instructions.
O4. Maintenance	Inadequate or routines not followed.
O5. Control and correction	Inadequate or routines not followed.
O6. Management of change and of design	Inadequate routines for planning, check and follow up.
O7. Competing operations	Different operations interfere with one other.
O8. Safety procedures	Missing, inadequate, disregarded.

T5   *Technical safety functions.* These are fulfilled by devices designed to reduce risk, such as machine guards, interlocks and various types of technical monitoring equipment. Examples of deviations include safeguards that are defective or inadequate and equipment that has been removed or disconnected.

Comments on human deviations

Since human errors are nearly always more complex by nature than technical failures (see Section 2.2), it is more difficult to provide a simple classification of human deviations. This means that certain types of human deviations, e.g. *forgetting something*, or *failing to take adequate safety precautions* can be placed in several categories. For this reason, flexible use should be made of this part of the checklist.

H1   *Operation/movement.* This applies to errors in the direct handling of material and equipment. They include "simple" errors of various types, e.g. slipping, falling over, missteps, etc.

H2   *Manoeuvring.* This category refers to the indirect handling of objects, using a machine or control system. Deviations include misreading of an indicator, error of judgement, choosing the wrong control button, moving an object in the wrong direction, etc.

H3   *Job procedure.* This applies where there is a normal task procedure. Deviations comprise various types of mistakes (errors of judgement), such as forgetting a step, doing subtasks in the wrong order, misinterpreting signals, etc. Also, a person may totally abandon the normal job procedure and work by means of stage-by-stage improvisation.

H4   *Personal task planning.* Most jobs allow several degrees of freedom, and provide scope for several types of deviations, in the form of both mistakes and violations. An unsuitable solution may be chosen for a variety of reasons, such as inadequate knowledge or a lack of instructions. Violations, such as breaching regulations and risk-taking, may have many different explanations (see Section 2.2). Not using personal protective equipment is a special case of this, and deserves special attention.

H5   *Problem solving.* This is a complex activity, and possibly too advanced to be analysed using the simple Deviation Analysis method. The reason for including this point in the checklist is to make an attempt to identify situations where there is room for a person to try to solve a problem in a hazardous way.

H6   *Communication.* This is an important component of many systems and job tasks. The category is used to identify situations where communications errors, either with another person or the system, can be hazardous. It covers missing information, misunderstanding, and misinterpretation. Also, erroneous or unclear messages can cause problems.

H7   *General.* This category is used to make an extra check in case there is an inconsistency in the system demands imposed on personnel. It can concern required physical or cognitive skills, or special knowledge needed for the job. The limitations of human beings may cause problems, either for themselves or for the functioning of the system.

Comments on organisational deviations

Not only human but also organisational failures are highly complex by nature. For this reason, flexible use should also be made of this part of the checklist. Organisational failures can be seen as the "root causes" of many technical and human failures—since it is decisions made during planning that generate the preconditions for other failures. One way of using the organisational part of the checklist is to treat important technical and/or human failures as a foundation. A check is then made concerning the extent to which organisational issues affect these failures.

O1   *Operational planning.* This is a general category which in principle also covers the points below. Planning can involve a variety of problems. It may simply be non-existent, or it may be incomplete or misguided.

O2   *Personnel planning.* This is a matter of having the right person in the right place. Problems include a lack of staff, staff without the required skills, and a lack of plans for training or recruitment.

O3   *Instruction and information.* Those who do the job must have adequate information on how to do it in the right way and according to plan. This might apply to manuals for equipment or to job descriptions for occasional tasks. Problems include a lack of instructions, and instructions that are inadequate, out-of-date or simply incorrect.

O4   *Maintenance.* In many cases, details of relevance to safety will be important. Problems include a lack of maintenance plans, plans that are not followed, important routine subsections that are missing, the unavailability of spare parts, and a way of working that is unsatisfactory.

O5   *Control and correction.* These are operations designed to ensure that equipment and activities function as planned. If they fail, the system should be returned to its "normal state", or plans should be modified as appropriate. Deficiencies in correction can cause a steady reduction in the safety of both technical and organisational functions.

O6 *Management of change and of design.* Inadequate routines for planning, check and follow up when systems are changed can result in reduced safety. For example, safety routines and responsibilities might be lost when a new organisation is set up.

O7 *Competing operations.* This category refers to situations where different operations can have a disruptive effect on one another. The operations may be quite independent, or they might compete for the same resources. For example, the number of cranes on a construction site is limited. If demand is great, a crane may not be available, prompting workers to resort to hazardous manual lifting.

O8 *Safety procedures.* These are designed to ensure that hazards are identified and controlled in accordance with the norms that prevail in the workplace. Safety management systems fall under this heading. The problem may be that safety activities are generally lacking or inadequate. Other deficiencies can include low or misguided priorities, unclear areas of responsibilities, poor routines and weak implementation.

Types of deviations

The checklist above is structured in accordance with system functions, but deviations can be of many different types. Table 7.3 provides a summary of types of deviations which can be used for different functions. Some specialised methods, such as HAZOP (Chapter 8) and Action Error Method (Section 11.3) employ similar checklists for types of deviations.

Table 7.3 Types of deviations.

None
Too little
Too much
Wrong type
Wrong order
Wrong place
Too late or early
Takes too long or too short a time

Selection of deviations

The number of deviations in a system can be very large. Which are most important and how extensively they should be examined will depend on the aim and level of ambition of the analysis. It may be that the analysis is limited to deviations of types 1 to 5 in Table 7.1. Alternatively, it might be restricted to one or some of the categories listed below. Deviations are divided into those that:

1. Can directly lead to injury.
2. Lead to the weakening or impairment of a safety function.
3. Require hazardous corrections in the course of production.
4. Seriously disturb production or make planning impossible.
5. Increase people's proneness to error (external disturbances).
6. Reduce control over the system.

Different levels of deviations

There is a hierarchy of types of deviations. Some directly lead to accidents, others to an increase in the probability that other deviations will arise, etc. Where there is a desire to work strictly with this hierarchy, the results from a Deviation Analysis can be used as raw material for more precise forms of analysis, such as Fault Tree Analysis (Chapter 8).

7.3 DEVIATION ANALYSIS PROCEDURE

General

Deviation Analysis is used to study a production system and the activities within it. The aim is to identify in advance deviations that can cause accidents or other problems. Usually the intention is to obtain a relatively broad picture of the hazards within a system. The analysis usually includes a stage where proposals to increase safety are generated.

The method can be applied on a small system, e.g. a workplace or an operation (such as repair work). A Deviation Analysis can also cover a large system, such as a whole factory. The principle is the same, but the approach to structuring and hazard identification will vary according to type of object.

Steps in procedure

Deviation analysis proceeds in a manner that is similar to Energy and Job Safety Analysis, and involves the same main stages. The analytic procedure is shown in Figure 7.1.

PREPARE

Preparations include defining how large a part of the system is to be covered by the analysis and specifying the operational conditions that are supposed to apply. At the same time, this determines what is not to be included in the analysis. A general piece of advice is not to be too restrictive in making this definition.

A second element in preparatory work is to ensure that the requisite information will be available during the course of the analysis (see Section 7.6). As an aid to analysis, a record sheet such as that shown in Table 7.6 can be used.

Deviation Analysis can be employed for problems other than accident risks. Before an analysis is embarked upon, a decision can be made on whether its scope should be widened to include other deviations, such as disturbances to production, poor product quality and damage to the environment (see also Section 4.4).

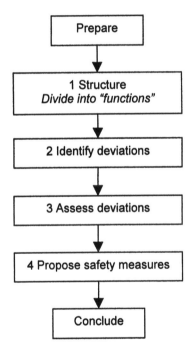

Figure 7.1   Main stages of procedure in Deviation Analysis.

1. STRUCTURE

The aim of structuring is to obtain a basis for the analysis. Its purpose is both to ensure that the entire system is covered and to divide it up into more elementary functions. The system is structured "functionally" on the basis of functions and related activities. The results of structuring can be represented as a flow chart and regarded as a model of the system.

The starting point is a description of operations. These are divided into blocks of an appropriate size. To cover general aspects, it may be a good idea to add a block denoted by such a heading as "General", "Planning" or "Organisation". This acts as a reminder to include organisational aspects when the examination is conducted.

Here are some examples of how a structure can be established:

- On a production line, a number of production steps are taken in sequence. The different links in the production chain can be followed, and these can be divided up into different sections.
- For a transport system, a classification can be made in accordance with the various types of conveyors used.
- A series of actions needs to be taken (a procedure). One example from everyday life is that of preparing a meal. A structure is obtained simply by listing the various actions required.

Structuring can be seen as creating a model of a system on the basis of what happens. Usually, there are certain self-evident main activities, and making the classification does not present any problem. On the other hand, there may be a number of subsidiary activities, which are not immediately apparent. Moreover, some of these may be hazardous, and must be included in the analysis. Examples of subsidiary activities include maintenance, the transportation of packaging material and the handling of waste.

When structuring is completed, a list of various sections or functions will have been obtained. These are then studied one at a time. Structuring is an important part of the analysis. It needs to be done with care, and a sufficient amount of time should be allowed for it.

2. IDENTIFY DEVIATIONS

The aim of this stage of the analysis is to find the most essential deviations. It is not possible to take up all conceivable deviations, as the total number can be very large. To start with, a consideration is made of how critical the section is for safety. If it appears to be essential to both safety and production, the section is studied more carefully. Otherwise, identification can be fairly quick.

For each section, an attempt is made to identify deviations that can lead to accidents or have other negative consequences. A good starting point may be to describe the purpose of the particular section under study. In searching for deviations, the checklist shown in Table 7.2 is used as an aid. For certain functions, e.g. for materials or procedures, the list of types of deviations shown in Table 7.3 can act as support.

The analysis can then proceed with the deviations that are assessed to be particularly important. For example, if a certain component is critical, it is possible to continue with the analysis by looking in particular at Maintenance (O4) and Control and Correction (O5).

Another example is where there are many possibilities for people to make errors in the handling of a machine and the skill of the operator is relevant to safety. Then, it can be important to look at Personnel Management (O2), Instruction and Information (O3), and perhaps also at Control and Correction (O4).

3. ASSESS DEVIATIONS

The next step is to assess the seriousness of the identified deviations. The principles for this are discussed in Chapter 4. The method itself does not prescribe what kind of assessment should be made.

In some cases, it is possible to obtain information on how frequent or serious deviations are, which can support the evaluation. Data can be obtained through interviews, from records of operations, or from notes about repairs. From accident investigations, data can be gathered on deviations that have involved a high level of risk. Especially if the company has an accident investigation method based on deviation investigation, see also Section 7.5.

4. PROPOSE SAFETY MEASURES

When the deviations have been assessed, an attempt is made to generate safety measures for those that are most important. At this stage, it is best to think as freely and creatively as possible, to develop a variety of ideas that can then be sifted through and modified. All the four types of safety measures referred to at the end of Section 7.1 should be systematically considered.

Eliminate the possibility that a certain deviation will arise is the first approach. This might mean a change in activity or device to remove the possibility. This type of measure is efficient but often difficult to implement.

To *reduce the probability* that a deviation will arise is more near at hand. Technical failures can be handled by better choice of components, maintenance procedures etc. (see also Section 2.1). Human errors might be improved with better man-machine interfaces, training, improved instruction manuals etc. (see also Section 2.2).

The third type of strategy is to *reduce the consequences* of a deviation. This might involve a technical solution, e.g. the installation of an interlock, or improving possibilities for the operator to recover the system if he should make a mistake in a sequence.

The fourth approach is to *support early identification of the deviation* and provide for plans on how it should be corrected in a safe and effective manner. This might be essential since many deviations cannot be avoided. A minimum requirement would be that operators know how to act when the deviation appears.

The checklist (Table 7.2) of system functions, particularly its organisational section, can also be used as an aid. Ideas are noted in the analysis record sheet. These are then sifted through, and what emerges is put together into a proposal containing a number of different safety measures.

CONCLUDE

The analysis is concluded by preparing a summary. This can contain accounts of the terms and conditions under which the analysis was conducted, the most important deviations and hazards, and the safety measures proposed.

7.4. EXAMPLES

One problem in giving practical examples is that a system needs to be described fairly extensively for it to be possible to understand how it operates and what can go awry. However, the two examples below provide rough outlines of how a Deviation Analysis proceeds.

Example 1: A conference

This example is taken from an environment that is not particularly hazardous. On the other hand, a conference involves a "procedure" with which many are familiar. Most people will also have come across various examples of the disturbances and deviations that can arise.

Suppose that an important conference is to be arranged. The conference organiser is very concerned that everything will go well, as several previous conferences have gone badly. If something goes wrong, the organiser may become the object of ridicule. A Deviation Analysis is conducted as a basis for planning.

The structure (Figure 7.2) contains the most important activities and starts with the block "Planning", where other functions and the purpose of the conference are included. The next step is "Invitation to conference", which involves attracting the interest of the correct target group, etc.

During the course of the analysis, the elements in the diagram can be divided up still further. Let us study the "Presentation" phase and divide it into some of its component parts. The first phase is "Speaking". The conference hall is large, so a public address system is needed. We assume that pictures or diagrams are to be shown. It is also important that the main aim of the presentation is included in the analysis. This is described as "Imparting knowledge".

Table 7.4 provides examples of what the analysis can generate for the subphases "Speaking" and "Showing pictures". There are a large number of possible deviations for these subphases alone. Obtaining an overall picture of possible disturbances and problems provides an opportunity for the better planning of the conference.

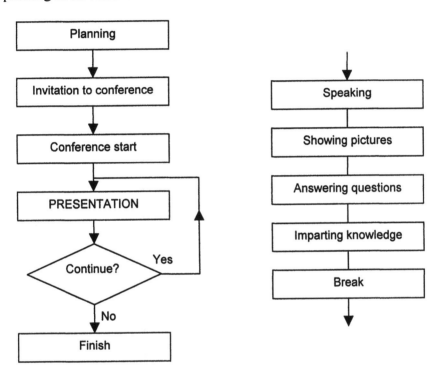

Figure 7.2 Schematic description of a conference, and a division of the PRESENTATION phase into blocks.

Example 2: Work with a computer-controlled lathe

Another example concerns work with a computer-controlled lathe of a fairly conventional design. We assume that production runs are short, so that the product to be manufactured is changed from time to time. This means that lathe settings are adjusted, and tools and computer programmes exchanged. There are a number of energies which can lead to serious injuries.

Our account covers structuring of the work (Figure 7.3) and examples of deviations (Table 7.5). Short summaries of assessment and finding safety measures are provided. Further, an extract from a record sheet showing how the analysis can be documented is presented (Table 7.6).

Table 7.4   *Deviations during conference presentation, coded in accordance*
with the types of deviations described in Table 7.2.

Function	Deviation	Code
SPEAKING		
Public address system	Does not work (see below)	T1
	Oscillates, howls	T1
	Failure of a component	T2
Managing the system	Faulty adjustment	H3
	Presenter (P) cannot cope with the microphone	H7, O3
General	Presentation takes too long time	H4, O5
	No checks on the sound equipment	O5
	Sound management not planned	O1
	No-one appointed to manage the sound	O2
	Inexperienced sound manager	O2
	Disturbance from drilling in an adjacent room	O7
SHOWING PICTURES		
Projector	Wrongly adjusted projector, blurred picture	T1
	Faulty projection, picture too small	T1
	Several other possible technical disturbances	T1
	Defective lamp or fuse	T2
Pictures/diagrams	Text too small	T3
	Poor translucence	T3
	Wrong picture order	T3
	Upside-down/back to front	T3
Visibility	Strong main lighting, picture not visible	T4
	Weak main lighting, P cannot read script	T4
	Picture obscured by person or object	T1
General	P cannot switch on the projector	H7, O3
	New picture in wrong direction	H2
	Failure to lower the blinds	O3
	Failure to turn on main lighting/spotlight	O3
	Failure to switch the lighting back on	O3
Planning	Lack of a projector	O1
	Lack of staff to give assistance when needed	02
	Inadequate instruction to staff	O3
	Projector not checked	O5
	No reserve materials	O4
	Equipment taken by other presenter	O7

1. STRUCTURE

General structuring generates six main phases (Figure 7.3.) which jointly comprise the procedure for manufacturing any one of the products. At the "Setting-up" phase, the operator will tool the lathe, change settings, read in the computer control programme, etc. At the "Testing settings" phase, the lathe is run for one programme sequence, and then stops. The operator makes certain checks on the settings and adjusts the control parameters if needed. When the entire job cycle has been tested, automatic operations can then be set in motion. The testing phase itself contains several subphases (also shown in Figure 7.3).

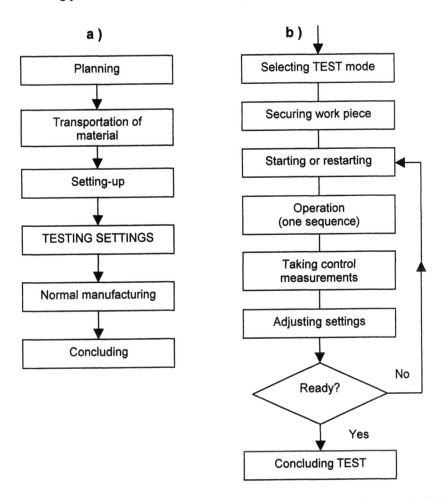

Figure 7.3   Working with a computer-controlled lathe. a) Production divided into blocks; b) the "Testing settings" phase.

2. IDENTIFY DEVIATIONS

Table 7.5 provides examples of different deviations that may occur in the course of "Setting-up the lathe" and "Testing settings". The examples come from a case study (Backström and Harms-Ringdahl, 1986), and all these deviations have in fact occurred. Some can lead to accidents, while others either cause defects in the finished product or mean that extra time must be taken to complete the work. Table 7.6 shows how a Deviation Analysis record sheet can be filled in.

There follow a few comments on some of the deviations. A "settings sheet" describes how the lathe should be set up, and which tools and control programmes should be used. The deviation "Select wrong settings sheet" means that the machine will be set up for the wrong type of manufacturing. Which settings sheet is to be used is listed in coded form on the works order. An error can be made by the operator or might have occurred earlier in the chain.

The first part of the testing phase is for the operator to initiate the' operational mode "TEST". The aim of selecting this mode is to run the job-cycle sequence at low speed. The most fundamental deviation is that this mode is not selected. The lathe will then go directly into production mode, and the tool settings, etc. will not be properly checked. This gives rise to a major risk for accidents or breakdowns.

There are a number of deviations which may have this effect. On the control panel, there is a button with the text "BLOCK DELETE". When the indicator light is on, the lathe is in automatic operating mode. If it is off, this means that "TEST" has been activated. Therefore, a defective lamp will mislead the operator. The control panel has a lot of buttons, labels and instructions, and there are many ways in which an operator can make a mistake.

3. ASSESS DEVIATIONS

The next step is to assess the seriousness of the identified deviations. In this case the approach shown in tables 4.3 and 4.4 is used, which means that both hazards to humans and production are directly assessed as being acceptable or unacceptable.

In Table 7.6 the assessments are shown. The first deviation was classified as S3 and P2, indicating that the situation was not acceptable from either a safety or a production perspective. The deviation "Incorrect measurement" was judged as S1 and P3, which meant that safety was not affected but it could be a serious production problem.

In this part of the production cycle, no deviation was judged as leading to health or environmental problems. This could have been demonstrated by writing a H0 and E0 in each row. But it is often practical just to make entries that indicate a non-acceptable situation.

Table 7.5   Some deviations when working with a computer-controlled lathe.

Activity	Example of Deviation
SETTING-UP	
Studying works order and settings sheet	Select wrong settings sheet Wrong number on the settings sheet
Fitting cutting tools and accessories	Select wrong tool Fit tool in wrong place Fit tool incorrectly Defective or worn-out tool
Removal of tools	Leave the old tool in place
Reading in programme	Error in loading procedure Data transmission failure Select wrong programme (for another product) Select out-of-date version of programme
TESTING SETTINGS	
General	Omit entire test procedure
Selecting TEST mode	Wrong indication of mode (lamp faulty) Press wrong button
Securing work piece	Inadequate fastening (technical or manual error)
Operation, one phase at a time	Excessive pressure of tool on work piece Work piece comes loose Speed of rotation too high (error in earlier installation, or technical failure) Work with safety hood open Operator puts head in machine to see better Unexpected machine movement (for the operator) Unwanted stop, e.g. with no READY signal Correction of disturbance in a hazardous way
Control measuring	Incorrect measurement
Adjusting settings	Wrong calculation Enter values incorrectly
Ready? (Continue test)	Finish the test before the entire cycle is completed (may depend on unclear information from the system)

4. PROPOSE SAFETY MEASURES

In the record sheet (Table 7.6) a number of measures are proposed. Several of these are aimed at reduction of probability of deviation. Several deviations can result in the work piece not being fastened securely. One conclusion was that a special investigation needs to be conducted of the various deviations that will cause a work piece to come loose. Moreover, as so many were found, Fault Tree Analysis was regarded as a suitable method for this.

7.5 ACCIDENT INVESTIGATION

About investigations

Investigation of accidents can give good understanding of how hazards can cause accidents in a workplace. Both research and practical experiences have indicated that accident investigations at companies can often be considerably improved. There are a number of methods for accident investigations, and a short account is given in Section 11.6.

The investigation of an accident differs in some respects from a safety analysis. An investigation covers just a limited part of the system and only some hazards. Moreover, the "choice" of hazards investigated is determined randomly by the accident that has occurred. However, the investigation can be transformed into a safety analysis by using the accident itself as a point of departure, and then carrying out a more thorough investigation of the system as a whole. For this, either Deviation Analysis or Management Oversight and Risk Tree (MORT as described in Section 11.5) can be employed.

About Deviation Investigation

This section describes a method called Deviation Investigation. It is closely related to Deviation Analysis, so the same checklist and a similar thinking can be employed. However, instead of searching for hypothetical deviations, an attempt is made to discover what preceded the accident in question. Near-accidents can also be investigated using the same method.

The aim of an investigation can vary. It might be to find out in detail what happened and the order in which events happened. Another goal could be to better understand the risks in the production system where the accident happened. This could then help in developing safety proposals. Our account highlights the latter aspect.

Compared with common investigations at companies, experience has shown that accident investigations conducted on the basis of deviations provide more information, and they also generate a greater number of proposals for safety measures.

Table 7.6 *Extract from a Deviation Analysis record sheet for working with a lathe.*
Both safety and production risks are evaluated using scales from tables 4.3 and 4.4.

Function / Part	Deviation	Hazard / Comments	Eval.*	Proposed measures
Testing settings / General	Omit the whole procedure	Work piece can come loose at full speed during later operation	S3, P2	Instruct the operators on the hazards of omitting the procedure. Extra interlock to avoid omission
/Selecting TEST mode	Wrong mode indication Press wrong button	Work piece can loosen at full speed / *Broken lamp gives a faulty indication*	S3, P2	Change design of indicator
/Securing work piece	Inadequate fastening	Work piece can come loose / *Many possible reasons; technical or manual failures*	S3, P2	Conduct a Fault Tree Analysis to summarise possible failures and errors
/Operation, one phase at a time	Work with safety hood open	Operator squeezed by tool or caught by rotating work piece / *Interlock installed*	S2	Improve interlock with e.g. dead man's grip when hood is open
	Disturbance, problem with sensors	Unwanted stop Squeezed / *If starts unexpectedly*	P1, S2	Develop safety correction routines Interlock with e.g. time out function for unwanted stops
/Control measuring	Incorrect measurement	Normally no danger, but the whole product batch can be destroyed	S1, P3	Instrument which is easier to use Better illumination
/Adjusting settings	Wrong calculation Enter values incorrectly	See above	S1, P3	Better training of operators and good calculation facilities
/Continue test	Finish too early	Work piece can loosen / *Possibly hazardous consequences at next stage*	S3, P2	Clearer indication from the system when test procedure is completed

* **Evaluation codes:** From Table 4.3 and 4.4:   S = Safety; H = Health; E = Environment; P = Production

0 = Negligible risk;   1 = Acceptable risk, no safety measure required;   2 = Safety measure recommended; 3 = Safety measure essential.

There are some arguments for using related methods for both accident investigation and safety analysis:

- Knowledge is obtained on which types of deviations have occurred during accidents at the company. This provides an aid for identifying relevant deviations and assessing their importance.
- Experience of investigation improves the skills needed for safety analysis, and vice-versa.

Deviation Investigation procedure

The investigation procedure is shown in Figure 7.4, and is very similar to the analysis of deviations in systems (see Figure 7.1 below).

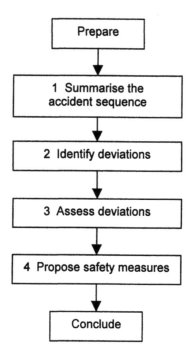

Figure 7.4   Main stages of procedure in a Deviation Investigation.

PREPARE

We assume a situation where a preliminary accident investigation has been conducted, and this needs to be supplemented. Having an investigatory team can be advantageous, especially in the assessment of identified deviations and the generation of ideas for improvements.

1. SUMMARISE THE ACCIDENT SEQUENCE

This is achieved using information from the preliminary investigation. The starting point is the accident event, which is then followed backwards in time (like rewinding a film in slow motion). Make a list of the deviations that arose.

2. IDENTIFY DEVIATIONS

This stage involves supplementing the list of deviations. In principle, the analysis continues backwards until everything in the system is "normal". The relevant information can be obtained from interviews with the injured person, job supervisor, people in the planning department, etc.

In the interviews, efforts are made to find new deviations, which have not previously been detected. There is also a need for follow-up information on those deviations which are already known. As an aid to identification, the checklist shown in Table 7.2 can be used. It needs to be reformulated in an appropriate manner to be of help in conducting the interviews. For example, ask:

- *Was the machine working normally? (T1)*
- *Had anything failed? (T2)*
- *Was there anything unusual about the materials being used? (T3)*

Deviations related to human errors need to be reformulated and expressed more concretely. For example, ask:

- *Were all the different parts of the job performed in the regular manner?*
- *Who planned the job?*
- *Should personal protective equipment have been used?*
- *Were there any misunderstandings involved?*

Questions on job supervision and planning may often be contentious, and are avoided in most investigations. But, organisational functions are important, and an investigation that does not consider them is incomplete. The checklist may make it easier to pose questions in such a way that they are not perceived as being loaded against any individual. Of course, the questions cannot have the exact form that they have on the checklist, for example:

- *Were planning procedures followed? (O1, M3 and M4)*
- *Was planning adequate? (O1)*
- *Were the job tasks of the injured person appropriate? (O2)*

On whether any technical equipment had been defective:

- *Why had the fault not been discovered before? (O5)*
- *Was the component covered by the maintenance programme? (O4)*

3. ASSESS DEVIATIONS

The next step is to assess the deviations. This can be done formally, as described in Chapter 4, or simply by selecting the most important deviations for further treatment. These may be the types which:

- Occur frequently.
- Are considered to be a problem.
- Are in breach of national regulations or company rules.

4. PROPOSE SAFETY MEASURES

Ideas for possible safety measures are based on the deviations selected. The measures are generated in the same way as when Deviation Analysis is used (Table 7.4). It is usually best to start with technical and individual deviations, and then investigate whether organisational conditions and routines can be improved. Try to think freely at the beginning.

The next step is to assess the ideas for safety measures and give them more concrete form. In doing this, it may be that new and better ideas will be generated to replace some of the old.

CONCLUDE

The investigation is concluded by making a summary recommendation for safety measures.

Some advice

We assume from the beginning that the purpose of the investigation is to generate proposals to raise the level of safety, not to find scapegoats. It is important to explain this to the people interviewed. It both facilitates discussion and makes it easier to obtain information.

It is not always possible to find out exactly what happened. There may have been different alternative sequences of events, or it can be uncertain whether a particular deviation really occurred. When searching for ideas for safety measures, these uncertainties need not be regarded as drawbacks. Rather, the situation is quite the opposite. If an accident can occur in several different ways, it is best that the safety measures cover all eventualities.

Example of an accident investigation

The accident occurred at a paper-rolling machine in a paper mill. The example is taken from a case study discussed in Section 15.6, and is the same type of machine used in the example of Job Safety Analysis (Section 5.3). The machine is used to cut up wide reels that come from a paper machine. The principle is that the paper from the large reel is wound up so as to pass through a system of rollers and reel cutters. The final result is a number of smaller reels.

The accident

On the occasion of the accident, the injured person (P) was about to pull a new paper reel through the machine. His hand slipped and he received a severe cut from a reel cutter. The preliminary accident report stated that the paper had become crumpled, and when P pulled the sheet of paper down to correct the fault, his hand slipped. The supplementary investigation identified four further deviations (5–8).

Figure 7.5. A new sheet of paper is drawn towards the reel cutters of a paper-rolling machine.

Deviations

A total of eight deviations were recorded in the summary. The categories from Table 7.1 are noted in brackets, but these will not usually appear in a summary.

1. P cut himself between thumb and forefinger when his hand was trapped in the gap between the upper and lower reel cutters (injury event).
2. P's hand slipped (H1).
3. P tried to correct the faulty paper feed, but in a hazardous way (H2 or H4).
4. The paper became crumpled (T3).

5.   The automatic equipment used to thread the paper through the machine functioned poorly (T1).
6.   The work team departed from accepted practice. They should have started afresh, and threaded a new sheet of paper through the machine (O1, possibly H3 or H5, depending on the circumstances).
7.   P was not aware of safe job procedure for threading paper through the machine (H7).
8.   P was an apprentice, working for the first day on normally scheduled job tasks. It was his third day at the machine (O2).

Deviations 3, 6, 7 and 8 were selected for further investigation. The investigation involved:

1.   Examination of the job introduction programme for new employees.
2.   Scrutiny of job instructions for the cutting machine.
3.   Checking whether the work team followed the instructions.
4.   Listing the disturbances that occurred at the cutting machine.

It was found that the company had an ambitious job introduction programme, but it was too much concerned with the company itself and too little with actual job tasks. Job instructions were relevant in principle but were phrased too generally. There were no instructions on what should be done when disturbances to operations occurred. The instructions available were followed but, as one person expressed it, "You couldn't really break the rules, anyway". The list of common operational disturbances was a long one.

Safety measures
On the basis of this accident, the following safety measures were proposed:

1.   The programme for the introduction of new employees should be modified. Greater emphasis should be placed on occupational hazards and how to act when operational disturbances occurred. The introduction of a sponsorship system (under which each new employee would be supervised by an experienced worker) and the appointment of special instructors were also recommended.
2.   The job instructions should be modified. A list of disturbances, showing what to do in each case, should be prepared.

These measures can be categorised in terms of reduction of probability of disturbance and "Planning for the identification and correction of deviations". Similar disturbances to production were involved in other accidents. For this reason, a number of technical modifications were later implemented so as to reduce the likelihood that disturbances occurred.

7.6 COMMENTS

General

Deviation Analysis is not a very commonly applied method in safety work. There are, however, two main reasons for including it in this book. The first is that experiences of applying the method show good results, especially when it is essential to consider technical, human and organisational factors simultaneously. The other is that the method has received a favourable reception from people who have learnt and employed it. It provides an additional tool and offers an opportunity to tackle problems that might be obvious, but which are difficult to handle systematically.

Deviation Analysis is more difficult than Energy Analysis and Job Safety Analysis. However, it is possible to select a degree of detail adapted to the skills of the analyst and to the type of system which is to be analysed. In most cases, the time to perform an analysis varies from half a day to a week.

The method is general by nature and is not only applicable to accidents. Many undesired events are preceded by deviations. For example, the principles are applied to interruptions to production, accidents leading to environmental harm, and fire and explosion hazards.

Planning and information

Most production systems are more complex than they seem at first sight. Information on the system and the problems that arise in use can be obtained from:

- Direct observation.
- Written descriptions and drawings.
- Interviews.
- Accident reports, operations records, etc.

A practical way of conducting an analysis is to form a study team. The team should contain people acquainted with technical functions, how the work is organised, and how it is carried out in practice. Then, there will be adequate access to information on the system and its problems. The number of conceivable deviations in a system can be considerable. In practice, there is often only time to study a limited number of deviations, which means that good capacity to discern and distinguish is needed.

Structuring

The aim of structuring in Deviation Analysis is to obtain a foundation for the identification of deviations. The result of structuring can be depicted in a flow chart and represents a model of the system. Development of such a representation of the system is an essential part of the analysis. Some theoretical aspects of modelling are discussed in Chapter 14.

People trying out the method for the first time often regard structuring as difficult. Using Energy Analysis or Job Safety Analysis, a structure can generally be found more or less immediately. Some of the difficulties are that:

1. There is seldom a ready-made structure available at a suitable level of detail. It is the job of the analyst to divide the system into functions.
2. Descriptions of system functions are often incomplete.
3. There are often several different ways of dividing up the system.

In addition, it can be hard to estimate in advance the degree of detail required. On some occasions, descriptions of general functions are enough; on others, details are needed. In the two examples given in Section 7.4, a general classification is made first, and then some of the categorised functions are broken down in greater detail.

This means that the task of describing and structuring a system can take longer than the identification part of the analysis. However, careful structuring can be of benefit in other applications than analysis. It can be used for the production of job instructions, or for general descriptions of the system as a whole.

In a way, the situation is simpler when a system is at the planning stage. Then, it is possible to work on the basis of the plans alone. In the case of systems that already exist, there are discrepancies between what is planned and what takes place in practice. Moreover, one is confronted with a much more complex reality. By its very nature, planning is incomplete, and systems are modified as they develop.

Identifying deviations

The checklist of deviations may seem a long one. But, the purpose of the list is to support their identification. It is not meant as a "template" or model that should be rigidly applied. In practice, there is no time to ponder over each item at great length. An experience is that users of Deviation Analysis tend not to make extensive use of the checklist after some period of familiarity with the method. It becomes natural for users to observe and search for deviations without it.

Information on deviations that have occurred in relation to previous accidents is valuable. A number of accident reports on the system, covering a period of several years, can be gathered together. Supplementary investigations can be conducted in conjunction with a person acquainted with the particular cases. This provides a list of deviations that have actually led to accidents, and which can be used at the identification stage. If it can be shown concretely that a specific deviation does lead to accidents, this can also increase the motivation of the study team.

One special comment should be made in relation to human deviations. The analysis is not principally designed to examine human behaviour in detail. The perspective is that human errors do occur, and that the consequences of errors need to be considered. If prevention of human error prove to be essential to system safety, measures can be sought in improved technical arrangements, man-machine interfaces, organisational planning, training, and so on. A more specialised method might also be employed (see Section 11.3).

Also with regard to organisational deviations, the method is not very "deep", and more specialised techniques might be needed. However, this can also be seen as an advantage. With detailed methods, the whole situation of making an analysis might be so overwhelming that an analysis never is made. Deviation Analysis offers here a compromise and gives reasonable detailed information.

Different levels of deviations

Obviously there are many types of deviations, and they can have different consequences. Some lead directly to a serious event, but depend in turn on other deviations having previously occurred. Others affect the likelihood of the appearance of further problems.

During the identification stage, a deviation can be recognised as one of the following:

a. A consequence of a deviation that has previously occurred.
b. One that leads to a deviation that is already known.

In such cases, reference can be made to the deviations already identified, allowing simplified descriptions. At the end of an analysis, the material can be structured so that related deviations are merged. If the number of deviations is large and their connections are complicated, a supplementary analysis could help. A suitable solution might be a Fault Tree Analysis or an Event Tree Analysis.

In the course of analysis, a list of different deviations is obtained. This can be seen as a one-dimensional description of a two-dimensional phenomenon. This may seem a cryptic remark, but should be clear to those familiar with Fault Tree Analysis, which can be regarded as a two-dimensional description (see Chapter 9).

8
Hazard and operability studies

8.1 PRINCIPLES

In the chemical process industry, there is often a potential for major accidents. There is also a tradition that hazards are identified and control measures taken. A number of authorities and organisations work with these issues. One method that has become well established in the chemical industry is HAZOP, an abbreviation for Hazard and Operability Studies. Extensive guidelines have been prepared on how the technique should be employed (CISHC, 1977; ILO, 1988; Taylor, 1994; Lees, 1996)

The basic idea behind HAZOP is that a systematic search is made for deviations that may have harmful consequences. The HAZOP technique is designed to stimulate the imagination of designers in a systematic manner, thus enabling them to identify conceivable hazards.

The system analysed is viewed as a technical process model. Hazards are defined as deviations that might cause damage, injury or other forms of loss.

HAZOP's characteristic elements are defined as follows.

INTENTION A specification of "intention" is made for each part of the installation to be analysed. The intention defines how that part of the installation is expected to work.

DEVIATION A search is made for deviations from intended ways of functioning that might lead to hazardous situations.

GUIDE WORD Guide words on a checklist are employed to uncover different types of deviations.

TEAM The analysis is conducted by a team, comprising people with a number of different specialisations.

The first section in this chapter provides an account of guide words, while the second describes the stages of procedure used for HAZOP. In Section 8.3 a simple example is provided. The chapter concludes with some comments and tips, principally obtained from original HAZOP specifications.

Guide words

One of the most characteristic features of HAZOP is the use made of "guide words". These are simple words or phrases applied to the "intention" of either a part of an installation or a process step. Guide words can be applied to:

- Materials.
- Unit operations.
- Layouts.

Table 8.1   Guide words in HAZOP.

Guide word	Meaning
NO or NOT	No part of the intention is achieved. Nothing else happens.
MORE	Quantitative increase, e.g. in flow rate or temperature.
LESS	Quantitative decrease.
AS WELL AS	Qualitative increase. The intention is fully achieved, plus some additional activity takes place, e.g. the transfer of additional material (in a conveyance system).
PART OF	Qualitative decrease. Only a part of the intention is achieved.
REVERSE	Logical opposite of intention, e.g. reverse direction of flow.
OTHER THAN	Complete substitution. No part of the original intention is achieved. Something quite different happens.

A simple example

The simple example that follows illustrates the use of guide words. The example refers to a liquid which is to be pumped into a pipe.

The first three guide words are immediately and easily understandable. NO means that nothing is pumped, MORE that more liquid than intended is pumped, LESS that less than intended is pumped; AS WELL AS means that something in addition to the intended pumping of the liquid takes place. AS WELL AS might refer to:

- The liquid containing some other component, e.g. from another pipe.
- The liquid also finding its way to a place other than that intended.
- A further activity taking place at the same time, e.g. the liquid starting to boil inside the pump.

The guide word PART OF means that the intention is only partially realised. If the part of the installation under study is designed to fulfil more than one objective, perhaps only one of these is met:

- A component of the liquid is missing.
- If the liquid is to be supplied to several places, only one of these receives its supply.

The guide word REVERSE denotes that the result is the opposite of what is intended. In the case of liquid, this might be that flow is in the reverse direction.

The guide word OTHER THAN means that no part of the original intention is realised. Instead, something quite different occurs. The guide word may also mean "elsewhere". In terms of the example, OTHER THAN might be due to:

- The pumping of a liquid other than the liquid intended.
- The liquid ending up somewhere other than intended.
- A change in the intended activity, e.g. that the liquid solidifies (or starts to boil) so that it cannot be pumped.

8.2 HAZOP PROCEDURE

The stages of procedure in HAZOP are extensively described in the literature referred to above. A rather simplified description is provided here. When using HAZOP, all stages are usually applied to each part of the process, taking parts one at a time. An example of the special record sheet used for HAZOP analysis is shown in Table 8.2.

PREPARE

The aim of the analysis has to be specified. It may be to examine the proposed design of an installation or to increase the safety of an existing plant by generating improved job instructions. The types of hazards to be considered can also be specified. These can concern hazards faced by people at the installation, product quality, or the influence of the plant on the surrounding environment.

A boundary for the analysis is set by specifying which parts of the installation and which processes are to be analysed. A team is appointed to conduct the analysis. As usual in safety analysis, preparation also involves the gathering of information and planning for the implementation of the study.

1. STRUCTURE
The installation is divided into different units. In the case of a continuous process, the division is into tanks, connecting pipes, etc. The analysis is then applied separately to each unit, one at a time.

2. SPECIFY INTENTION
The intention of each part to be analysed is defined. This specifies how it is envisaged that the part will function. If the designer participates, he or she can provide an explanation. Otherwise, it will be the person most familiar with the installation.

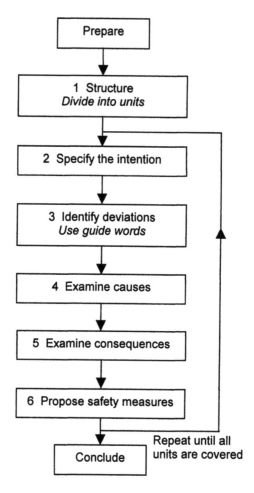

Figure 8.1   Main stages of procedure in HAZOP.

3. IDENTIFY DEVIATIONS
Using the guide words, an effort is made to find deviations from the specified intention. The guide words are applied one at a time.

4. EXAMINE CAUSES
For each significant deviation, an attempt is made to find conceivable causes or reasons for its occurrence.

5. EXAMINE CONSEQUENCES
The consequences of the deviations are examined. The possible seriousness of these should also be assessed. Matters of assessment and grading of consequences are not taken up in the HAZOP manuals. It is possible to use the types of assessment discussed for the other methods.

6. PROPOSE SAFETY MEASURES
For deviations that may have serious consequences, an effort is made to find control measures. This stage is not always included in descriptions of HAZOP, but it may be a natural part of any analysis.

To obtain ideas for improvements, the same strategy as in Deviation Analysis might be used (Section 7.3). Safety measures may apply to:

- Changing the process (raw materials, mixture, preparation, etc.).
- Changing process parameters (temperature, pressure, etc.).
- Changing the design of the physical environment (premises, etc.).
- Changing routines.

REPEAT THE PROCEDURE
When analysis of a unit of the installation is completed, this is marked on the drawing. The next unit is then analysed, and the procedure continues until the entire installation has been covered.

CONCLUDE
The analysis is concluded by preparing a summary, but further follow-up might be needed. This might include liaising with those responsible for control measures, further development of safety proposals, etc.

8.3 EXAMPLE

The HAZOP analysis illustrated in Figure 8.2 is a simplified and amended version of an example originally presented by the UK Chemical Industry and Safety Council (CISHC, 1977). It concerns a plant where the substances A and B react with each other to form a new substance C. If there is more B than A, there may be an explosion. This is a highly simplified example. It is not specified whether a continuous or batch process is involved, how the quantities of A and B are controlled, etc.

Analysis

We begin with the pipe, including the pump, that conveys Material A to the tank. The first step is to formulate the INTENTION for this part of the equipment. Its aim is to convey a specific amount of A to the reaction tank. In addition, the pumping of A is to be completed before B is pumped over.

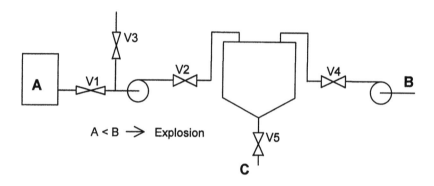

Figure 8.2   Schematic description of an installation.

We apply the first guide word, NO or NOT. The deviation is that no A is conveyed. The consequence of the deviation is serious, and involves the risk of explosion. Possible causes of this are sought for, and it is easy to come up with several conceivable explanations:

1. The tank containing A is empty.
2. One of the pipe's two valves (V1 or V2) is closed.
3. The pump is blocked, e.g. with frozen liquid.
4. The pump does not work, for one of a variety of possible reasons. The motor might not be switched on, there might be no power supply to the motor, the pump might have failed.
5. The pipe is broken.

The next guide word is MORE. The deviation means that too much A is conveyed. Reasons for this might be that:

1. The pump has too high a capacity.
2. The opening of the control valve is too large.

Consequences will not be as serious this time. But C can be contaminated by too much A, and the tank can be overfilled.

The third guide word is LESS, meaning that too little A is conveyed. The consequence may be serious. Reasons for this might be that:

1. One of the valves is partially closed.
2. The pipe is partially blocked.
3. The pump is generating a low flow, or is operating for a shorter time than intended.

The fourth guide word is AS WELL AS. The deviation is that A is conveyed, but that something else happens. Examples of such deviations are that:

1. A further component is pumped through the pipe, which might be due to Valve V3 being open, resulting in another liquid or gas entering the flow. Or that there are contaminants in the tank.

2. A is pumped to another place as well as to the tank. This might result from a leak in the connecting pipe.

3. Another activity is taking place which competes with the pumping. Would it be possible for A to boil in the pump?

The consequence of all these different deviations is that *too little A* is conveyed, meaning the risk of explosion.

The fifth guide word is PART OF. The deviation is that just a part of the intention is fulfilled. It might be that a component of A is missing, although this appears not to be possible in this case.

The sixth guide word is REVERSE. This would mean that liquid is conveyed from the reaction tank to the container for material A. The consequence can be serious. Conceivable deviations include:

1. The pump is operating in reverse. This would occur if the power supply was wrongly connected to the motor.
2. Liquid is running backwards from the reaction tank or the connecting pipe due to gravity.

The seventh guide word OTHER THAN means that no part of the original intention is fulfilled. Instead, something quite different occurs. Some examples of such deviations are:

1. A liquid other than the intended liquid is pumped.
2. The liquid finds it way to some other place.
3. There is a change in the intended activity. It might be that the liquid solidifies or starts to boil, so that it cannot be pumped.

Record sheet

Table 8.2 shows a part of the analysis summarised on a HAZOP record sheet. Such sheets can be designed in quite different ways. For example, the first column with guide words can be regarded as unnecessary. If assessments of identified hazards are made, a column for writing the assessment score is needed.

8.4 COMMENTS

Hazard and operability studies are well established, and a large amount of collective experience has been accumulated. The reference literature (CISHC, 1977; Kletz, 1983; ILO, 1988; Taylor, 1994; Lees, 1996) offers plenty of advice on how analyses should be used, planned and conducted. Many of these pieces of advice reflect the more general viewpoints on safety analysis expressed in Chapter 13. Several of them are based on experience of just this method, and a fairly extensive account is therefore also given in this section.

The basis for a safe installation lies in the application of an accepted, well-established technique for which a long period of experience is available. Various specialists can contribute their own specialised knowledge. In addition, there are different norms and directives from the authorities that must be applied.

All installations have their own special features, and hazards can manifest themselves in various ways. This is why there is a need to conduct an investigation of each installation to find specific faults and items that have been neglected in its design.

It should have emerged from this account that HAZOP does not provide an automatic means for the discovery of hazards and the generation of safety measures. The results obtained, as is the case with most analytical methods, are dependent on creative thinking and good knowledge. *Good results come from the application of a systematic approach, utilisation of the guide words, and the formation of a team with suitable membership.*

When is HAZOP used?

HAZOP can be used in different situations:

- At the planning stage, before detailed design and construction decisions are made.
- Before system start.
- For an existing installation.

Table 8.2   Record sheet for a HAZOP analysis.

Guide word	Deviation	Possible causes	Consequences	Proposed measures
NO, NOT	No A	Tank containing A is empty V1 or V2 closed Pump does not work The pipe is broken	Not enough A, explosion	Indicator for low level Monitoring of flow
MORE	Too much A	Pump has too high a capacity Opening of V1 or V2 is too large	C contaminated by A Tank overfilled	Indicator for high level Monitoring of flow
LESS	Not enough A	V1, V2 or pipe is partially blocked Pump gives low flow, or runs for too short a time	Not enough A, explosion	See above
AS WELL AS	Other substance	V3 open, air is sucked in	Not enough A, explosion	Flow monitoring based on weight
PART OF	–			
REVERSE	Liquid pumped backwards	Wrong connection to motor	Not enough A, explosion A is contaminated	Flow monitoring
OTHER THAN	A boils in pump	Temperature too high	Not enough A, explosion	Temperature (and flow) monitoring

At the planning stage

The greatest benefit is obtained if an analysis is conducted in conjunction with the design of the installation. The optimal point in time is when decisions are being made on how the plant is to be constructed and detailed design documentation is ready (design freeze). Drawings that are sufficiently detailed for the analysis are then available.

A HAZOP analysis takes time. In the case of a large installation, it can be a matter of several months, even if several teams work in parallel. It is possible either to embark on the construction work and accept the risk of changes or to wait until the analyses are ready. But construction schedules must allow time for this.

Before system start

There can be a point in conducting a HAZOP analysis even when the installation has been nearly completed and when instructions for users have already been prepared. The reasons why an analysis is justified at this stage are:

- Important changes have been made.
- Operating instructions are critical to safety.
- The new installation is similar to one that already exists. The changes primarily affect the process and not the equipment.

For an existing installation

An installation where safety was adequate at the time when operations were started may deteriorate over the years. A series of changes may have meant that different types of hazards have arisen. This particularly applies if safety issues were not carefully considered when the changes were made. It may also be that sufficient attention was not paid to safety at the design stage, or that requirements for operational safety have become stricter over time.

Information

The literature on the method stresses the importance of the availability of a sufficiently detailed documentary base for the analysis to be conducted. This means, among other things, that a HAZOP analysis cannot be conducted at too early a stage of the planning of an installation. A start on the study can only be made when detailed documentation is available.

Drawings and instructions must be up-to-date and correct. Drawings often need to be updated, which can require a substantial amount of effort. In the case of existing installations, it is often found that information is incorrect.

The study team

Guides to how HAZOP should be conducted stress the importance of working as a team. This applies to team composition, skills and attitude. HAZOP is no substitute for knowledge and experience. If the team lacks either of these, the results of a study will be unusable. It is important that members of the team have a positive and constructive attitude towards their task. Success depends on the ability of participants to think constructively and with imagination. Members must be selected with care, and motivation must be promoted.

The team should not be too large, containing a maximum of seven members. A study requires members with different specialised forms of technical expertise, with knowledge of the process, measuring and control techniques, etc. The team must have sufficient knowledge on how the installation is designed to function. Thus, for reasons of efficiency alone, it is important to be able to obtain answers directly, and avoid having to guess or to obtain information from outside sources.

If the team contains members who have the authority to make direct decisions on changes, this makes the study more effective. If the installation is being designed or constructed by an outside supplier, representatives of both user and supplier should participate in the study. The work requires continuity, so members should only be replaced in cases of emergency.

The role of the team leader is important. He or she must be familiar with the HAZOP method, capable of leading the discussions, and able to ensure that the schedule for analysis is followed. The task of the leader also involves producing the documentation needed for the study. It is sufficient for the leader alone to have thorough knowledge of the method. Other members can participate without extensive training. A training period of between one hour and two days has been mentioned, depending on the level of ambition of the study. The leader must ensure that proceedings at meetings are efficient and agendas kept to. There must not be so many delays that the members get bored with the analysis. The leader summarises results when each unit in the study has been completed. He/she also marks the drawing after, for example, a pipeline is ready.

Time taken by the analysis

For most installations, a HAZOP analysis is time-consuming. For this reason, proper scheduling is required. The average period of time required by an analysis is 10–15 minutes, either per component or per activity covered by a job instruction. This means one to three hours for each main unit, e.g. a reactor with several connecting pipelines. For analysis to be effective, study meetings lasting three hours at most are recommended. Moreover, these meetings should not take place more than twice or three times a week.

Thus, if the object to be analysed is a large one, careful planning is required. Planning involves the following:

- Finding time for the entire object.
- Getting through the meetings in a reasonable amount of time.
- Having the necessary information material available at meetings.
- Ensuring that time is available for control measures and follow-up activities decided upon at the meetings.

To complete the analysis in a reasonable time, several teams working in parallel may be required. One of the team leaders should then adopt the role of co-ordinator.

Safety measures

The finding of safety solutions can be conceived of in terms of two extremes. In practice, there will be a compromise between the two:

- A solution is produced after each source of risk (hazard) is discovered.
- No solutions are produced until after all the guide words have been applied.

The team is often predominantly, or even exclusively, composed of technicians. In such cases, it must be remembered that not all problems are solved by making technical changes.

The follow-up of measures and other such activities are important. Who is responsible is noted on the record sheet. If the analysis is conducted at the planning stage, or in the case of a new installation, there should be a readiness to make changes. In the case of an existing system, measures are needed so that the system will function better than before.

Analysis of batch production

The HAZOP literature already referred to contains supplementary advice for the study of installations where batch production takes place. In addition to drawings of the plant, information is needed on the sequence in which the procedure is carried out. It may be either automatically or manually controlled. The information material may consist of job descriptions, flow sheets, etc. A summary description of the settings of valves, etc. may be needed for different situations that can arise

The analysis can be structured so as to follow job procedure rather than different parts of an installation. The same guide words as before are employed, although these can be re-formulated as appropriate. For example, EARLIER, LATER and WRONG ORDER may be employed for time or job sequences. When applied in this way, the method is similar in certain respects to Deviation Analysis.

Miscellaneous

Taylor (1979, 1994) has suggested a variant of HAZOP in which the emphasis is on physical variables. The analysis is then based on a checklist that covers temperature, pressure, etc., and a simplified set of guide words is applied to these.

Some companies wish to receive detailed documentation of the analyses. This is sensible in itself, but involves a significant amount of extra work. According to Kletz (1983), it is seldom that such information is made use of afterwards.

If a lot of changes are made after a HAZOP study, a new round of analyses may be required. The additional study would be designed to discover whether new problems had been introduced by the changes already implemented.

Experience has shown that problems of start-up, close-down, etc. are often neglected by over-specialised design groups working in isolation. Sometimes, the guide word MISCELLANEOUS is employed to capture deviations or problems that have not been identified using the other guide words. The category is primarily designed to cover occasional activities that can lead to problems. Examples include starting up and closing down the plant, inspection, testing, repairs, cleaning, etc. This guide word does not have a natural place in HAZOP, but can be valuable for the detection of further problems.

9
Fault Tree Analysis

9.1 INTRODUCTION

A fault tree is a graphical representation of logical combinations of causes of a defined undesired event or state. Examples of types of final events are an explosion, failure of equipment, the release of toxic gas and an interruption to production.

Fault Tree Analysis (FTA) is perhaps the best known method employed in safety analysis. It started to be used in the 1960s. The method is of greatest value for complicated technical systems where a functional failure can have serious consequences, and also where considerable resources can be allocated for hazard analysis. The method is relatively difficult and is generally used by specialists. There is an extensive literature on the method (e.g. IEC, 1990; Kumamoto and Henley, 1996; Lees, 1996), and a number of computer programmes are available to aid the design of fault trees and make calculations.

It can be questioned whether the method is appropriate for common safety work outside high-risk sectors. But a general knowledge of Fault Tree Analysis is useful even for those who will not use the method directly. The purpose of the description given here is to acquaint the reader with the method and provide a basis on which simpler kinds of fault trees can be generated. However, probabilistic estimates, which form an important area of application in Fault Tree Analysis, are only briefly referred to. To some extent, the traditional focus of Fault Tree Analysis has been extended. As it now encompasses more than just technical factors, human actions and the taking of control measures are also considered.

Some of the advantages of Fault Tree Analysis are:

1. It is an aid for identifying risks in complex systems.
2. It makes it possible to focus on one fault at a time without losing an overall perspective.
3. It provides an overview on how faults can lead to serious consequences.
4. For those with a certain familiarity with the analysis, it is possible to understand the results relatively quickly.
5. It provides an opportunity to make probabilistic estimates.

Some of its disadvantages are:

1. It is a relatively detailed and, in general, time-consuming method.
2. It requires expertise and training.
3. It can provide an illusion of high accuracy. Its results appear advanced and, when probabilistic analyses are conducted, these can be presented in the form of a single value. But, as with most methods, there are many possible sources of error.
4. It cannot be applied mechanically and does not guarantee that all faults are detected. In general, different analysts will produce a variety of different trees. But a tree can have different forms and still have the same content.
5. Its implementation generally requires detailed documentary material to be available.

9.2 PRINCIPLES AND SYMBOLS

In fault trees, events and logic gates are basic concepts. In Fault Tree Analysis an either/or approach is adopted. Either an event occurs or it does not. An event statement can then be designated as "true" or "false". This can also be expressed in terms of the logical values "1" and "0", meaning that binary logic and Boolean algebra can be applied.

This is both a strength and a weakness. The approach has the advantage that faults in complex systems can be described in a simple manner. Its weakness is that many differences of degree that exist in reality cannot be taken account of by the analysis.

In designing a fault tree, a set of symbols is used. The set has a number of variants, and only a limited selection of the symbols is taken up here. Symbols in Fault Tree Analysis are of two kinds—gates and events. The most important are shown in Figure 9.1.

The first three symbols refer to "events" that describe a fault of some kind. They may be events in a strict sense, i.e. something that happens, but may also refer to a faulty state, e.g. a component that has failed. They might, therefore, be better described as "failure events".

The conditional symbol is used to show how normal conditions or events can also affect the system. Sometimes, the symbol is used in combination with a special gate called INHIBIT. The transfer symbol is used to divide a tree into several smaller parts.

The AND and OR gates are used to provide logical connections between the various events. A somewhat more extensive description is provided in Section 9.4.

Symbol	Designation	Function
 ◯	Basic event	Basic event or failure.
 ◇	Undeveloped event	Cases are not developed.
▭	Event	Event resulting from more basic events.
▭	Conditional event	Event that can occur normally.
 ⌓	AND gate	Output event C occurs only if all input events (A and B) occur simultaneously.
 ⌂	OR gate	Output event C occurs if any one of the input events occurs.
 △	Transfer symbol	Indicates that the tree is developed further at another place.

Figure 9.1   Symbols used in Fault Tree Analysis.

Example of a fault tree

The appearance of a fault tree may be illustrated by a simple example. A lamp is connected into a circuit, as shown in Figure 9.2. A power supply feeds the lamp, and there is a battery to provide reserve power in case the power supply fails. A fault tree is wanted to analyse the case where the lamp does not light when switched on.

The top event is that the lamp does not light. This is because there is no current through the lamp. In turn, this may be due to the lamp being faulty or there being no power supply to the lamp. The power feed will fail if both the power unit and the battery fail to operate (AND gate).

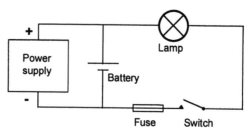

Figure 9.2 Example of a lamp circuit.

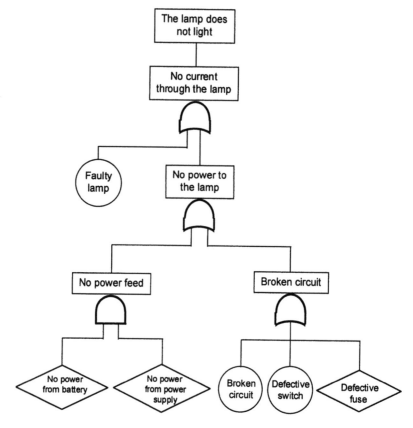

Figure 9.3 Fault tree for a lamp circuit.

The tree contains three basic events, and there are also three "undeveloped events". That the fuse is defective may be due to ageing or some other factor. But it might also have been overloaded—as a result, for example, of a temporary short-circuit. It should be possible to develop this further. Similarly, it should be possible to investigate why power is not coming from the battery or the power supply.

9.3 FAULT TREE ANALYSIS PROCEDURE

A Fault Tree Analysis cannot be conducted in such a direct manner as the analyses described in previous chapters. Constructing a tree is as much an art as a straight-forward building operation. Success much depends on the ability of the person performing the analysis. For this reason, it is difficult to provide a universal and clear description of how one should set about it.

Probabilistic applications usually involve four main steps—system definition, fault tree construction, qualitative evaluation and quantitative evaluation. One variant, with emphasis placed on the construction stage, is shown in Figure 9.4.

PREPARE
As is usual with safety analysis, pre-conditions need to be defined before the analysis itself can be conducted. Constructing a fault tree involves detailed analysis and may require an extensive set of assumptions. These may apply, for example, to the boundaries of the system under study and the operational conditions that are supposed to prevail. Assumptions may also be needed on which types of faults might occur and which should be excluded from the analysis.

1. SELECT TOP EVENT
The first step is to select the undesired event to be analysed. This should be carefully defined. If a top event is too broadly defined, it can probably be divided into several different events. A separate fault tree can then be constructed for each case.

2. SUM UP KNOWN CAUSES
When constructing a fault tree, existing knowledge of faulty states and failure events should be utilised. It facilitates the analysis if a preliminary examination of failures that may arise is conducted. Alternatively, the results of a deviation or HAZOP analysis can be used. This material can be used to construct part of the tree.

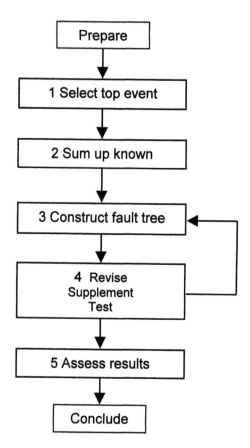

Figure 9.4   Main stages of procedure in Fault Tree Analysis.

After this step, a list of faults that might contribute to the occurrence of the top event will have been obtained. Generally the list will not be complete, but it will still be of major assistance in constructing the tree.

3. CONSTRUCT FAULT TREE
Construction of the tree begins with the top event. The first step is to consider whether it can occur in more than one independent way. If so, the system has to be divided up using OR gates. The analysis continues by moving downwards, searching for more basic causes. Some of these can be obtained from the preliminary list referred to immediately above.

4. REVISE, SUPPLEMENT AND TEST

Construction is a trial-and-error process. Progress towards a better and more complete tree is made in stages. A number of rules of thumb for carrying out this work are provided below.

It is hard to know exactly when the tree should be considered complete. No important causes of failure should be omitted. A first check is to see whether all the points on the preliminary list have been covered.

5. ASSESS RESULTS

The completed tree is then assessed, and conclusions are drawn. Depending on the purpose of the analysis, a number of different steps can be included at this stage. Some of these are discussed more extensively in Section 9.4.

- *Direct evaluation of the result.* The tree provides a compressed picture of the different ways in which the top event might occur. It also provides a picture of the barriers (safety features) that exist. A check can be made if some failures can directly lead to the occurrence of the top event.
- *Preparation of a list of minimum cut sets.* As shown in Section 9.4, a cut set is a collection of basic events, which together can give rise to the top event. A minimum cut set is one which does not contain a further cut set within itself.
- *Ranking of minimum cut sets.* Combinations of failures to which special attention should be paid can be evaluated and ranked on the basis of the minimum cut sets.
- *Estimation of probabilities* is the "classical" application of a fault tree. If information on probabilities for bottom events is available, or if these can be estimated, the probability of the occurrence of the top event can be calculated from the list of minimum cut sets.

CONCLUDE

The analysis is concluded with a summary, which gives information about assumptions. It is not enough with just the tree, which might be difficult to understand and interpret. Probably a number of conclusions can be made based on the analysis.

Rules of thumb

In constructing a fault tree, the rules of thumb shown in Table 9.1 can be utilized. Rules 1 – 7 are applied in the course of constructing the tree. Rules 8 –10 are used from time to time to test whether the tree has a valid logical structure. The list of rules is partly based on the account provided by Henley and Kumamoto (1981). A further source is the author's experience of problems encountered by beginners when they first embark on Fault Tree Analysis.

Table 9.1   Rules of thumb for constructing and testing a fault tree.

1.   Work with concrete events and states. These should be denoted in the form of event statements that can be either true or false.
2.   Develop an event into a further event that is more concrete and basic.
3.   Divide an event into more elementary events (OR gate).
4.   Identify causes that need to interact for the event under study to occur (AND gate.
5.   Link the triggering event to the absence of a safety function (AND gate).
6.   Create subgroups frequently, preferably dividing into pairs.
7.   Place a heading above every gate.
8.   Do not make tacit assumptions, and do not let preconceived opinions (non-explicit assumptions) control the analysis.
9.   Think logically and in terms of structure. Do not confuse cause and effect.
10. Test the logic of the tree from time to time during construction. Start from events at the bottom of the tree and suppose that these occur. What will the consequences be?

Figure 9.5 provides examples of how some of these rules of thumb might be used. Rules 6 and 7 have been merged into one. There is one OR gate with four inputs. This has been divided up into several gates, but the tree still has the identical logical function. If any of the basic events occur, then Event Statement A is true. Such a subdivision makes for more systematic analysis. The disadvantage is that the diagram takes up more space.

The example of Rule 8 shows a line of thought that has been neglected. That the machine starts unexpectedly will only lead to an accident if a person is directly in the danger zone.

The example of merged rules 9 and 10 shows a case of confusion of cause and effect. Suppose that the motor operates for too long. This does not lead to current flowing for a long time. In this case, the sequential error is obvious, but in more complicated contexts it is easy to perform such logical somersaults. Remember that causes start at the bottom!

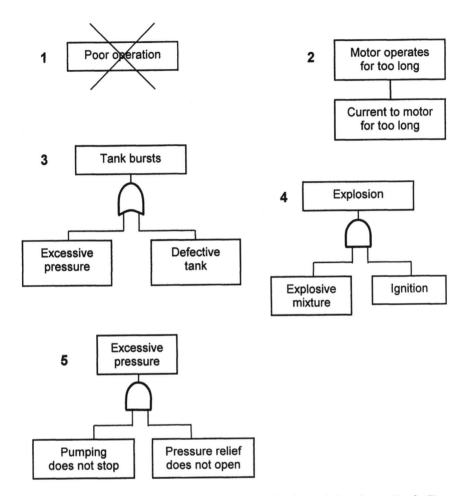

Figure 9.5a Examples of the application of rules of thumb in Fault Tree Analysis (part one).

9.4 MORE ON FAULT TREE ANALYSIS

General

This section takes up a number of additional themes related to Fault Tree Analysis, such as types of symbols, other kinds of trees and forms of evaluation.

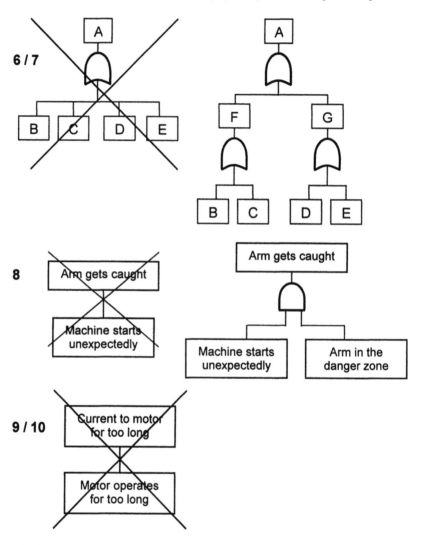

Figure 9.5b Examples of the application of rules of thumb in Fault Tree Analysis (continued, part two).

Traditional applications of Fault Tree Analysis involve the making of quantitative estimates. In such cases the main stages are as follows:

1. Definition of the system.
2. Construction of the fault tree.
3. Qualitative evaluation.
4. Quantitative estimation.

Defining the system is often the most difficult part of the analysis (Lees, 1996). It is important to be well acquainted with the system and how it functions. Its physical boundaries need to be specified, and further conditions may need to be defined. This can apply to:

1. The top event.
2. Initial conditions.
3. Failures that might be supposed to occur.
4. Excluded events.

Only a few accounts of how a fault tree is constructed in practice are available. Perhaps the most extensive is that of Henley and Kumamoto (1981), several of whose ideas have been utilised in preparing the list of rules of thumb presented above. The authors have also further developed guidelines for fault-tree design (Kumamoto and Henley, 1996). The construction of a fault tree is a combination of art and science. Two analysts will not construct identical trees. (But note that this also applies to safety analysis in general as soon as a trivial level is passed.)

Large numbers of computer programmes of different types are available as aids for the construction of fault trees. For a beginner, it seems best to start by constructing a tree by hand. Otherwise, there is a substantial risk that the task of managing the programme will be predominant and analytical thought neglected.

On the use of logical symbols

The most important symbols are shown in Figure 9.1 above, but they can be presented in alternative ways. The functions can also be expressed in the form of a set of logical expressions or as truth tables. Figure 9.6 shows how these different forms of presentation are related.

Let us start with the logical variables A and B and their corresponding event statements, which may be "true" or "false". If A is "false", it is given the value 0; if A is "true", it obtains the value 1. Examples of what A might represent are *motor switched on,* or *safety device removed.*

AND and OR gates may have an arbitrary number of inputs. For AND gates all input statements must be true for the output statement to be true. For OR gates it is enough for just one of several input statements to be true for the output statement to be true.

A further function has been added, i.e. the negation (NOT). This is not used in Fault Tree Analysis, but still demands some attention. Negation statements take the following form: if A is true, then Z is false.

A truth table shows how a logical function depends on the input variables. This can be explained through the examples given in Figure 9.6:

- NOT (Z) is a function of a variable (A). When A = 0, Z = 1. For A = 1, Z = 0.
- The AND gate (X) has two inputs. For A = 1 and B = 1, X = 1. For other combinations of A and B, X = 0.

Probability functions have been included in Figure 9.6 (see also Section 2.1). Fault Tree Analysis is frequently employed to provide a basis for probabilistic calculations, and it is appropriate to introduce some basic formulas here. The probability that A will occur within a certain time interval is denoted as p(A). The probability that A will not occur is 1 – p(A). For the formulas for p(X) and p(Y) given in Figure 9.6 to be applicable, it is assumed that A and B occur independently of each other.

Function	AND	OR	NOT
Symbol	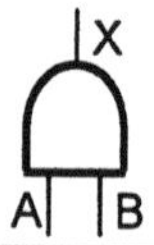 $\mid$X $\bigcap$ A$\mid$ $\mid$B	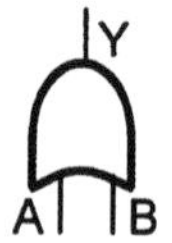 $\mid$Y $\bigcap$ A$\mid$ $\mid$B	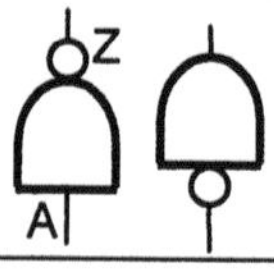 $\bigcirc$Z $\bigcap$ A$\mid$
Alternative symbol	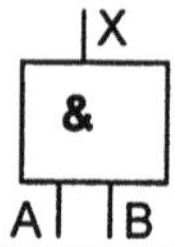 $\mid$X & A$\mid$ $\mid$B	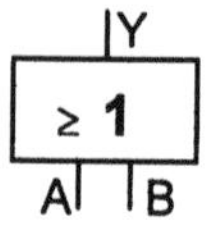 $\mid$Y ≥ **1** A$\mid$ $\mid$B	
Function denotation	X = A B X = A & B (X = A ∩ B)	Y = A + B (Y = A ∪ B)	Z = A' (Z = $\bar{A}$)
Truth table	A 0 1 B 0 0 0 1 0 1	A 0 1 B 0 0 1 1 1 1	A 0 1 1 0 Z
Probability	p(X) = p(AB) = p(A) p(B)	p(Y) = p(A + B) = p(A) + p(B) – p(A)p(B)	p(Z) 0 1 – p(A)

Figure 9.6   Different ways of describing logical relationships.

Things that break

That an installation breaks is because its load is greater than its strength. For a more extensive description, see O'Connor (1981). Normally, installations are designed and constructed so that there is a margin between lowest strength and highest load. If a failure occurs, this may be because the margin is too narrow. Let us take a bridge as an example.

This is illustrated in Figure 9.7. The load is not constant but varies over time. Sometimes there are a lot of vehicles on the bridge, at other times there are only a few or none. The curve on the left shows the probability (p) of the bridge being exposed to a certain load.

Nor is the strength constant. The bridge can rust, or extreme cold may mean that it is weaker at certain times. Even if a large number of identical bridges have been built, it is not certain that all have equal strength. Construction errors or material defects can arise. For this reason, a curve is needed, which shows the probability that the bridge will cope with a certain force.

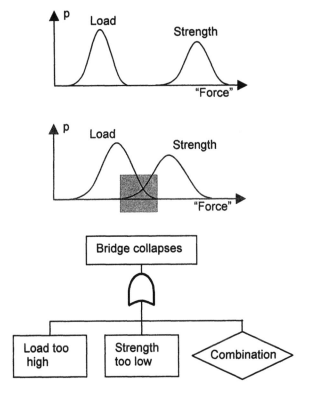

Figure 9.7   Relationship between probability and strength. The shadowed area shows that the construction might collapse.

In the case of the upper of the two curves, there is a margin between load and strength. How large the safety margin should be is decided at the design stage. For example, the relation between maximum permissible load and strength might be set at a factor of ten. The lower curve shows a situation where the margin is insufficient. Sooner or later the bridge will collapse.

A fault tree can be marked to denote that load is high relative to a certain specified value, or that strength is lower than this value. A combination of these faults can also arise. Sometimes it can be difficult to distinguish between the two cases.

Component failures are often classified as primary failures, secondary failures and command faults (linked using OR gates). Primary failures occur during normal operating conditions, e.g. from the effects of natural ageing. Secondary failures occur when a component is exposed to conditions for which it is not designed. Command faults refer to functions where the component does work but where its function cannot be fulfilled, e.g. as a result of signals that are faulty or absent.

Other types of trees

Relationships between different functions can be described by various types of trees, not just fault trees. It is easy to confuse different types of trees. Three are shown in Figure 9.8. An organisation can be illustrated in the form of a tree, and a "hierarchical" tree can show the order of relations between departments. Such trees can also be used to describe technical systems.

A classification into subgroups or classes can also be illustrated by a tree. The word taxonomy is used to describe the classes created when there is a strict classification. Such a tree is not a fault tree, but can form part of one. It can be used to distinguish between different events that may have the same final result.

A success tree can be used to describe what is required for an installation to work. Such trees are also described as logic-flow or function diagrams. They are the opposite of fault trees, which show what is required for something not to function.

Relationship between fault and success trees

There is a close relationship between a fault tree and a success tree. The lighting of a lamp is described as a fault tree in Figure 9.3, and as a success tree in Figure 9.8. In the fault tree, the functions are negative; they concern what is defective. For the bulb to light, everything must work (AND gates). For the bulb not to light, it is enough for there to be just one fault (OR gate).

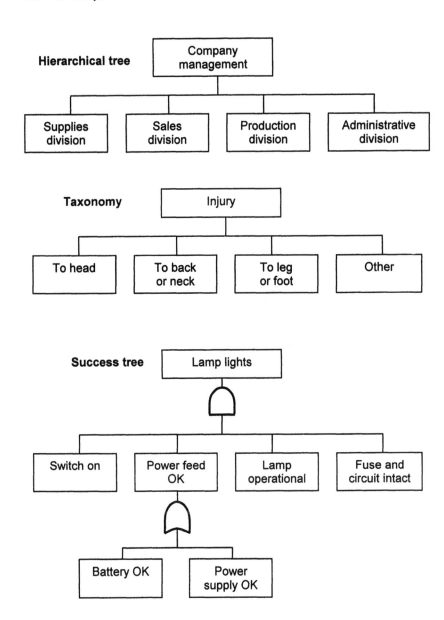

Figure 9.8 Examples of tree diagrams.

There is a general way of transforming a success tree into a fault tree.
De Morgan's theorem states that:

$$(AB)' = A' + B', \text{ and} \tag{9.1}$$

$$(A + B)' = A'B' \tag{9.2}$$

The theorem can be proved by setting up truth tables for the right and left
sides of the equations. It can then be observed that the tables are identical in all
positions. The theorem can be generalised to encompass more than two
variables.

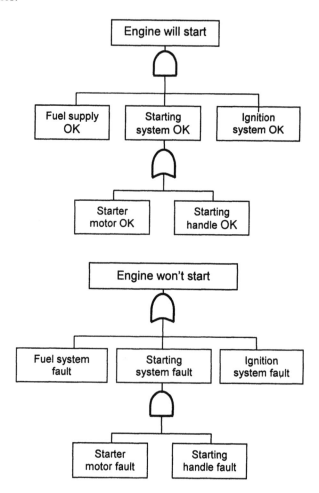

Figure 9.9   Example of transforming a success tree into a fault tree.

This can be expressed in words as a simple rule. A success tree can be transformed into a fault tree by:

1. Negating all statements, i.e. writing the opposite.
2. Transforming all AND gates into OR gates.
3. Transforming all OR gates into AND gates.

It is a mistake, however, to consider that a fault tree can be constructed simply by transforming a success tree in accordance with this rule. A fault tree must go more deeply into the "negative" side, taking account of different types of faults in depth. Moreover, it is rare that the top event will be of the type "System does not work". Nevertheless, the way of thinking described above may be of assistance. It can be applied to either a complete system or a subsystem. An example of a transformation is provided in Figure 9.9.

Simple preliminary evaluation

A fault tree can be used as a basis for making probabilistic estimates, but it is also possible to draw direct conclusions from studying the tree. Some of the questions raised in such an evaluation are as follows:

*   Are there only OR gates in the tree? This might mean that the tree is of poor quality or perhaps not a fault tree at all. Or it might mean that the system is highly vulnerable, since all faults will lead to an accident.
*   Are there basic events that directly lead to the top event? This means that a single basic failure will lead to an accident.
*   Are the system's safety features included in the tree? (These appear as AND gates.)
*   Can the level of safety be increased? The tree can give ideas where a safety barrier may be useful, e.g. showing when a single failure can cause an accident. Symbolically, such barriers will appear as AND gates.
*   Are assumptions clearly specified? Or are important assumptions implicit, e.g. that electrical power will be supplied the whole time?

Can "common cause" failures be a serious problem? This means that faults which are supposed to be independent are in fact triggered by the same event. Examples include loss of electric power, and that several human errors arise in sequence as a result of poor instructions or the misinterpretation of a situation.

Ranking of minimum cut sets

As a basis for further evaluation, the tree is often divided up into minimum cut sets. A cut set is a collection of basic events that can give rise to the top event.

A minimum cut set is one that does not contain a further cut set within itself. In the example from Figure 9.9 the minimum cut sets contain:

- "Fuel system fault"
- "Ignition system fault"
- "Starter motor fault" AND "Starting handle fault".

In the case of simpler trees, a division into cut sets can be carried out by hand. However, there are a number of computer programmes available that can provide assistance both in identifying cut sets and in making probability calculations.

A qualitative assessment can be made to reveal which basic events make the greatest contribution to the occurrence of the top event (e.g. Brown and Ball, 1980). The ranking of minimum cut sets is based on two separate factors. The first is the number of basic events that are included. If there is just one, the set has greater significance than where two or more are involved. The second concerns types of faults where human errors are seen as most likely. In many cases, it as advantageous also to include "organisational errors" in the human-error category. Table 9.2 gives suggestion for the ranking of cut sets according to their importance.

Table 9.2   Ranking of importance of cut sets in a fault tree.

Combination	Comment
HE	One human error (HE)
AC	One active component failure (AC)
PC	One passive component failure (PC)
HE & HE	Combination of two human errors
HE & AC	
HE & PC	
AC & AC	
AC & PC	
PC & PC	
HE & HE & HE	Three independent human errors
HE & HE & AC	Etc.

Probabilistic estimates

A fault tree can be used for estimating the probability of the occurrence of the top event. Estimates of probabilities for all the bottom events of the tree are needed for this. The reader wishing to go further into probabilistic methods in Fault Tree Analysis should refer to the more specialised literature (e.g.

Kumamoto and Henley, 1996; Lees, 1996). There are various computer programmes available, which will help with the calculations. The greatest general problem is to find failure data of sufficient quality on the various components of the system.

An approximation of the probability of the occurrence of the top event is derived by summing the probabilities of the minimum cut sets. This presupposes that these probabilities are low.

An alternative calculation procedure involves working directly from the bottom events in the tree, moving upwards stage-by-stage and applying formulas for the AND and OR gates (see Figure 9.6). This provides a clearer picture of which types of faults make the greatest contribution to the occurrence of the top event. The correctness of the result depends on two conditions: that bottom event failures are independent of one another, and that each bottom event appears in only one place within the tree.

Probabilistic estimates have a number of benefits. For example, they enable solutions to be compared and provide assistance in setting control priorities. However, there are also a number of difficulties. Lees (1996) summarises some of the problems involved in using Fault Tree Analysis as a tool for estimation:

- The fault tree may be incomplete; there is no guarantee that all faults and all logical relationships will be included.
- Data on probabilities may be lacking or incomplete.
- Estimates for systems with low failure probabilities are difficult to verify.

9.5 EXAMPLE

System description

Figure 9.10 shows a sketch of a chemical processing plant. In the tank, two chemicals react with each other over a period of 10 hours and at a temperature of 125°C. When the reaction is complete, the contents are tapped off into drums through the opening of a valve.

The two chemical ingredients are pumped over from two other tanks. The volumes pumped are read off on two special instruments. The contents of the tank are heated by a coil controlled by a relay. The temperature rises at a rate of approximately 2°C per minute when the heating device is connected, and falls at roughly the same rate when it is off.

The temperature is measured using a sensor. The signal from the sensor is linked to the relay and forms a part of the temperature control circuit. If the temperature is lower than required, the relay switches the heating on. If the temperature is too high, the heating is turned off.

The signal from the sensor is also connected to an alarm that is activated if the temperature exceeds 150°C. If the alarm sounds, the operator is supposed to switch off the power feed manually.

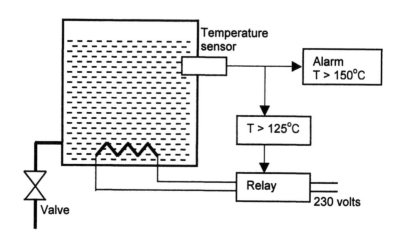

Figure 9.10   Reaction tank with heating and alarm facility.

Preparing the analysis

A fault tree is required for the event that poisonous gas is formed. It is based on the system description above. The level of accuracy of the description is low, but it is sufficient for a preliminary analysis. We assume that a proposal as specified in the sketch has been made, and that the analysis is designed to evaluate this proposal.

Selecting the top event

The proposed top event *Poisonous gas is formed* can be used directly.

Summing-up known causes

No previous investigation has been made. (A HAZOP or Deviation Analysis might have been a suitable start, which would give a number of deviations to include in the tree.) It can be seen directly that there are some possible faults, which may be hazardous. Let us look at some examples:

- Sensor out of order, giving a low temperature reading.
- Temperature circuit controlling the relay function does not turn off the power.
- Alarm circuit failure.

"Sensor out of order" will lead to a hazardous situation. Other failures can arise, but these mean that the temperature will be too low. Such failure events are not included in the fault tree.

Constructing the fault tree

We start with the top event and see that it is caused by the heating element operating for too long. The analysis might then continue in several different ways. We decide to divide the system into two parts, i.e. before and after measurement of the temperature. Two failure events are then relevant:

1. The information on temperature is incorrect at the output of the monitoring circuit.
2. The heater cut-out does not work (despite receipt of a correct signal).

We now have the upper part of the fault tree (Figure 9.11), and go on to consider temperature measurement error (Figure 9.12). This might occur due to a fault in the measuring circuit, but no details of the circuit are available. So, we draw in a rhombus to mark an undeveloped event. That the measured temperature at the sensor is too low may have a large number of different causes. The most obvious is that the level of the liquid is below the sensor.

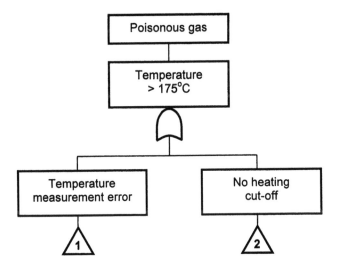

Figure 9.11 The upper part of the fault tree.

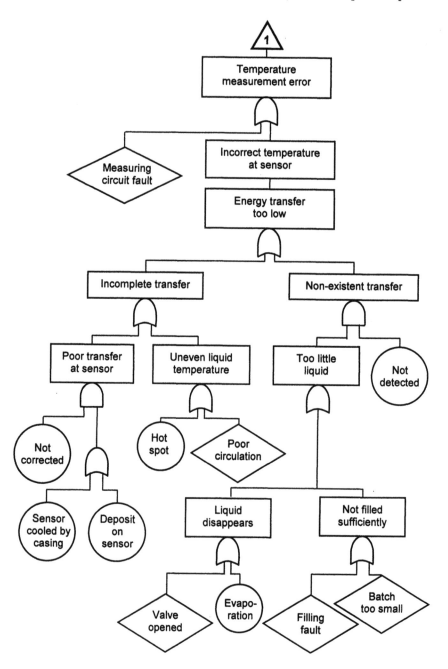

Figure 9.12 Fault tree for temperature measurement error.

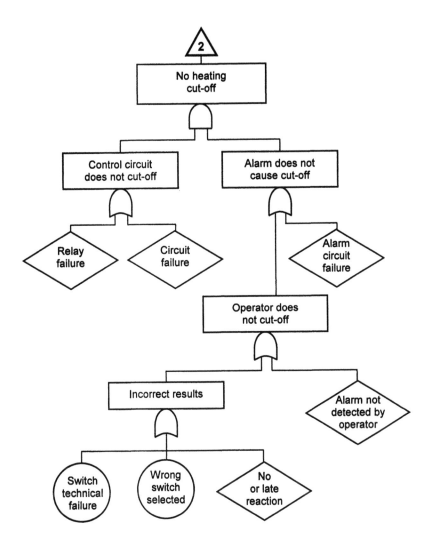

Figure 9.13 Fault tree for cutting off heating device.

The part of the tree that shows how a failure to cut off the heating device may arise is shown in Figure 9.13. We do not have much information on its technical design, so the fault tree is rather small. That the alarm does not interrupt the system may have several causes. We do not know how the alarm will be installed, or which people should take action when it sounds. Despite this, some types of general failures can be marked.

Two protective actions have been assumed. One is that the operator has been instructed to make a report if deposits appear on the sensor. The other is that the operator should report if the liquid does not reach the sensor. These are treated as assumptions and included in the fault tree. But, as the tree describes failure events, these protective actions are marked as *"Not corrected"* and *"Not detected"*. For these safety functions to be of any real value, they must be part of a routine check. Just for this reason, they can be considered as rather weak.

A general comment on the design of the tree is that there is too little space on the tree itself for adequate descriptions of failure events. The significance of each event will have to be made clear by its context and by surrounding events. An indexed list of the events on the tree may be needed.

Revising, supplementing and testing

The tree has been developed though a process of revision and supplementation. It would require too much space to show how this takes place. One simple test is to check for the inclusion of the "known causes" produced at the beginning of the analysis.

Assessing the results

The first assessment is to inspect the tree, looking especially for safety functions (i.e. AND gates). It can be seen that the tree includes three failures that lead directly to the formation of poisonous gas. One is a measuring circuit fault. The other two are both related to *"Uneven liquid temperature"*, which can be due to *"Poor circulation"* or a *"Hot spot"* on the heating element.

A simple list of minimum cut sets in rank order is made directly from the tree. We adopt the principles described in the previous section using the factors Human error (HE), Active component failure (AC), and Passive component failure (PC).

The tree contains three purely technical failures, which on their own can lead to the occurrence of the top event. There are also three combinations of two human errors, which lead to an accident. Moreover, there are a number of other combinations of two faults, which lead to the top event.

Although the tree appears large enough, it is still not complete. For example, the situation where the tank has not been emptied completely is neglected. If the heating is switched on by mistake, the temperature will be too high. Such a situation might to some extent be fitted into *"Liquid disappears"*, but it is still not fully covered. This is because the analysis only treated the system during continuous operation.

Table 9.3.   Ranking of combination of events in a fault tree.

Combination of events	Comments
AC	Single failure of active component.
1. Measuring circuit fault	Not developed, but a separate tree could investigate this further.
2. Poor circulation	As above, several reasons are possible.
3. Hot spot	
HE & HE	Double human error
4. Valve opened & Not detected	The operator releases some liquid, and does not realise that it may be hazardous.
5. Filling fault & Not detected	Too small an amount of liquid has been inserted from the beginning.
6. Batch too small & Not detected.	Too small a batch of the chemicals is ordered. That this can be dangerous is not obvious.
HE & AC	Human error and failure of active component
7. Deposit on the sensor & Not corrected	Deposit on the sensor results in too low temperature reading. And the deposit is not removed.
8. Sensor cooled by casing & Not corrected	Sensor is not adequately insulated from its casing, and this gives a too low temperature reading. Moreover, it is not detected and corrected.
9. Evaporation & Not detected	The quantity of liquid diminishes, e.g. through evaporation. Also, this is not detected.
10 –15. Control circuit does not cut out & Operator does not cut off	Six combinations are possible. Not listed here. (Three human errors and two technical.)
AC & PC	Failure of active and passive component.
16 –19. Failures in control and alarm circuits.	Four combinations of technical failures in the control and alarm circuits are possible. Not listed here.

Conclusions

In retrospect it can be said that the range of the analysis was too narrow. The greatest risks may lie in the heating being on when the operating procedure is started or stopped.

The conclusions about the technical system that can be drawn from the tree are as follows:

1. Three (or nine) single-failure events may lead directly to the occurrence of the top event. The safety level has to be regarded as unacceptable.
2. The installation requires radical re-design and an increased level of safety.
3. The new design and construction proposal will require analysis. Both what precedes and what follows the heating phase must be considered.

Comments

The tree differs in several respects from a conventional tree. Above all, this applies to the bottom events. In many cases, they are not failures in that they refer to a technical fault or have ceased to function. Rather, they can be characterised as departures from what have been designed or planned. Thus, they are equivalent to "deviations" in the sense employed in Deviation Analysis and HAZOP. To go further and attempt to estimate probabilities, the tree must be more stringently constructed. Some of the bottom events must be defined more precisely, and the relations between them further specified.

There are ten rhombuses as bottom events indicating that they could be developed further. To do that more technical and organisational information is needed. Eight items at the bottom have been ringed, marking faults that are fundamental. However, it is not obvious which bottom symbols should be chosen, and several of the ringed sections could be further analysed in new trees.

Some of the basic events are classified as "Human errors", such as "Not corrected" or "Not detected". These could also be seen as organisational failures, e.g. if there is no maintenance procedure for checking for a deposit on the sensor.

In total there are 18 bottom events in the tree. But this number could easily be doubled if a more thorough analysis was conducted. Even in a case as simple as this, a fault tree can become very large.

9.6 COMMENTS

Learning Fault Tree Analysis

Fault Tree Analysis is the most difficult of the methods presented so far. Its disadvantages and advantages have been described in Section 9.1. After proper training it is not difficult to conduct an analysis, but it requires effort and may take a long time. If hazards with major consequences are to be studied, using Fault Tree Analysis is justified.

The length of time it takes to learn the method obviously depends on level of ambition and previous knowledge. It may be fairly easy for electronic engineers and computer programmers who are trained in the handling of logical circuits and functions.

Some problems

One problem is that a tree may be large and require a large amount of time to develop. A second problem is that the analysis may have too great a focus on technical failures. Human and organisational factors may be neglected. For this reason, immediate concentration on technical failures alone should be avoided. If technical aspects prove to be extremely important, they can be studied more deeply at a later stage of the analysis.

A simplified picture

A fault tree is a simplification of reality (in many senses). One concerns the adoption of the either/or approach, another the setting-up of strict logical connections between events. This means that some information must be excluded. Perhaps these problems are greatest if there is a desire to go beyond a conventional technical application.

Checking existing trees

It may be that a fault tree is already available, and the analyst has the task of evaluating and interpreting it. A fault tree can be seductive, encouraging the belief that it is complete and has been well thought through. What might have been overlooked? Some of the questions to address are as follows:

- What assumptions about and simplifications to the system have been made?
- Are only technical failures included?
- Does the tree accord with "rules of thumb"? (The ones given in Table 9.1 are in no way generally accepted or applied, but they do provide some sort of measure of quality.)
- Are there only OR gates? (If all failures will lead to the occurrence of the top event, there are grounds for wondering whether the tree is correct.)

A variety of applications

The fault tree methodology can be applied for a variety of purposes, which are not mutually exclusive. They include:

- Making probabilistic estimates.
- Identifying alternative chains of events that might lead to an accident.
- Compiling and providing a logical summary of results from other analyses, such as Deviation Analysis. A tree can be used to make a transition from a (one-dimensional) list of deviations to a representation of their logical connections and relations.
- Organising information from the investigation of an accident that has occurred. It can provide a means for examining the results obtained.

In many cases it can be fruitful to construct a tree with an accident that has occurred as the top event. Some of the reasons for this are as follows:

- Motivation for making use of the results is stronger. When a serious accident has occurred, the top event is no longer an abstract one. Instead, it is something of direct relevance.
- The compact description offered by a fault tree places emphasis on the overall picture and not on particular details.
- In some cases, there is uncertainty over which deviations and failures actually occurred. Instead of viewing uncertainty as a weakness, it can be regarded as offering a starting point for identifying alternative ways in which accidents may occur.
- Broadening the range of investigation and searching for deeper explanations become part of a natural process.

10
Analysis of safety functions

10.1 INTRODUCTION

This chapter takes up approaches and methods for describing and analysing safety characteristics of a system. Most methods for safety analysis are oriented towards identification of failures and problems, and how these can be corrected. There are some negative aspects to this, for example:

- The correction procedure takes time.
- Changes to hardware and procedures may be expensive.
- It might be difficult to judge the final result, especially if a number of safety measures are added one by one.

Another approach is directly to study and evaluate the safety features of a system. This has some potential advantages, for example:

- Safety functions (both technical and organisational) can be properly designed from the beginning.
- A comprehensive description of the safety characteristics of a system is possible.
- Support is given to design specification and clarification of the interfaces between systems and responsibilities.
- Whether safety functions have sufficient efficiency and coverage is evaluated. (Is the system safe enough?)

A study (Harms-Ringdahl, 1999) has been performed to see which methods and principles are available to support such an approach. Of special interest was the possibility to apply this on more common workplaces where it is essential to adopt a rather simple methodology. Conclusions of the study were that:

- There is varying terminology, sometimes with poorly defined terms. This can be expected to cause confusion in many situations.
- The field of safety and safety functions is less theoretically developed than expected.
- There is a potential for development of both theory and methodology.

- There are a number of methods for analysing safety features. However, they appear to be too complicated for application at common workplaces.
- The concept of "safety function" is worthy of greater attention.

As a consequence of the last conclusion, the method "Safety Function Analysis" has recently been developed. The aim was to obtain a method that directly focuses on safety features, but not in too complicated a manner. The final section of this chapter describes the method, and includes an example.

10.2 SAFETY IN DIFFERENT SECTORS

Nuclear power sector

Safety within the nuclear power area is well documented in numerous reports. A summary of basic safety concepts in the nuclear power sector is provided by International Nuclear Safety Advisory Group (INSAG, 1988). Twelve fundamental safety principles are discussed, and they are divided into three main groups. A compressed overview is given in Table 10.1. In the report, a set of 50 "specific safety principles" is also discussed. In a later report (INSAG, 1996) the characteristics of "defence in depth" in nuclear safety are further described.

Table 10.1   Summary of general safety principles (from INSAG, 1988).

Main groups	Safety principle
Safety management	Safety culture Responsibility of the operating organisation Regulatory control and independent verification
Defence in depth	Defence in depth Accident prevention Accident mitigation
Technical principles	Proven engineering practices Quality assurance Human factors Safety assessment and verification Radiation protection Operating experience and safety research

Chemical industry sector

The chemical industry also has a long tradition of systematic safety work. A comprehensive overview of safety principles is provided in the "Guidelines for Safe Automation of Chemical Industries" (CCPS, 1993). It describes both general aspects, and also safety in connection with automated safety and process control systems.

A fundamental term employed is *"protection layer"*, although this is not explicitly defined. It "typically involves special process designs, process equipment, administrative procedures, the basic process control system and/or planned responses to imminent adverse process conditions; and these responses may be either automated or initiated by human actions".

A figure entitled *"Protection layers"* displays eight levels. These are arranged in order of how they are activated in the case of an escalating accident:

1. Process design.
2. Basic controls, process alarms and operator's supervision.
3. Critical alarm, operator's supervision and manual intervention.
4. Automatic safety interlock system.
5. Physical protection (relief devices).
6. Physical protection (containment devices).
7. Plant emergency response.
8. Community emergency response.

Automation of technical systems

An essential part of the guideline (CCPS, 1993) concerns automation aspects and control systems. A design philosophy for *safety interlock systems* is encapsulated in ten distinct points.

There is also a general standard called *"Functional safety: safety related systems"* from the International Electrotechnical Commission (IEC, 1998). The standard covers the aspects that need to be addressed when electronic systems are used to carry out safety functions. It is extensive and contains seven parts.

The scope is to set out a generic approach, one that is independent of application. Examples are given from process and manufacturing industries, transportation, and the medical arena. The standard is mainly concerned with safety to persons. A number of basic terms are employed in the standard:

- *Safety-related system* implements the required safety functions necessary to achieve a safe state for the equipment under control. (A person could be part of a safety-related system.)
- *Functional safety* is the ability of a safety-related system to carry out the actions necessary to achieve a safety state for the equipment under control.

- *Safety integrity* is the probability of a safety-related system satisfactorily performing the required safety-related functions under all the stated conditions within a stated period of time.

Energy barriers

Energy models have been used for a long time, and they usually involve technical as well as organisational aspects of barriers. These are described in greater detail in Chapter 5 on Energy Analysis. Energy barriers are also central to Management Oversight and Risk Tree (MORT). See Section 11.5.

In MORT (Johnson, 1980) barriers are defined as physical and procedural measures to direct energy in wanted channels and control unwanted release. A categorisation is also made, which is similar to the one used in Energy Analysis (Table 5.3).

The energy concept and energy barriers fit quite naturally into other applications. One example is the "Safety Barrier Diagram" method described in Section 10.3.

Other aspects

A general concept is "defence", which can represent several types of safety features. It has been discussed in detail by Reason (1990, 1997). In simple terms, defences shall prevent that hazards cause losses. Such defences can combine in several layers, but can be weakened by different kinds of problems. These can be caused by active failures, e.g. unsafe acts. Latent conditions, such as poor design, reduce the "strength" of the defences. A combination of active failures, latent conditions, and local circumstances might cause an accident to occur.

Organisational aspects are highly relevant to the modelling of safety characteristics. An interesting example is a framework for modelling safety management systems (Hale *et al.*, 1997). Safety management is seen as a set of problem solving activities at different levels of abstraction, and risks are modelled as deviations from normal or desired process. Safety tasks are modelled using the Structured Analysis and Design Technique (SADT).

There are also a number of alternative approaches to analysing and describing safety characteristics (e.g. Kecklund *et al.*, 1995; Hollnagel, 1999). The terminology for describing barriers and safety functions varies quite considerably. There are also several ways in which they can be classified (for overviews see Harms-Ringdahl, 1999; Hollnagel, 1999). "Safety function" is a common concept, but no clear definitions have been found in the literature. Definitions and applications are further discussed in sections 10.4 and 10.5.

10.3 METHODS FOR ANALYSIS OF BARRIERS AND SAFETY

There are a number of methods that can be used for analysis of barriers and safety functions. One group of methods are those that are specially designed for this purpose. There are also other methods, which more or less can be adapted for such a purpose.

A list and brief descriptions of the methods belonging to these two groups is given below.

Specifically designed methods

- *AEB.* The Accident Evolution and Barrier Function Method (Svenson, 1991; 2000) can be used for analysis of accidents and incidents (see Section 11.6).
- *MORT.* Management Oversight and Risk Tree (Johnson, 1980) can be used for analysis of systems and accidents (see Section 11.6).
- *SADT.* Structured Analysis and Design Technique (Hale *et al.*, 1997) can be used for analysis of safety management systems (see Section 11.6).
- *Safety Barrier Diagrams* (Taylor *et al.*, 1989; Taylor, 1994) offer a way to present and analyse barriers to accidents (see Section 10.3).
- *Safety Function Analysis* (Harms-Ringdahl, 2000) can be used for analysis of safety characteristics of a system (see Section 10.3).

Adaptable methods

- *Energy Analysis.* Barriers are a fundamental part of the method (see Chapter 5)
- *Event Tree Analysis.* One common application (Rouhiainen, 1993) is to check a safety function to see whether or not an event gives rise to damage (see Section 11.2)
- *Fault Tree Analysis* can show how barriers and safety features might prevent an accident (see Chapter 9).

10.4 SAFETY BARRIER DIAGRAMS

An approach called "Safety Barrier Diagrams" is a way of presenting and analysing barriers to accidents (Taylor *et al.,*1989; Taylor, 1994). The term "safety barrier" is used to describe a safety device or other measure that can prevent, reduce, or stop a given accident sequence. A more detailed definition (Taylor, 1994) is:

- Any wall, shield, switch, bolt, interlock, software or operational check which is intended to prevent a signal or activation from reaching a place where it can cause an accident.
- A mechanical barrier which can prevent external influences from causing an accident.
- A mechanical barrier which can prevent a release of energy or poison from having adverse consequences.
- Distance from the source of hazard.

A "safety configuration" is defined as a combination of safety barriers. Figure 10.1 illustrates the basic structure of a Barrier Diagram.

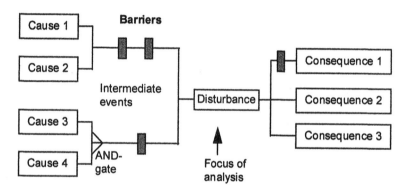

Figure 10.1 Principles of the Barrier Diagram approach (adapted from Taylor et al., 1989)

A safety diagram is constructed with the disturbance as centre point. Possible consequences are shown to the right, and causes and initial events to the left. If two or more causes coincide an AND gate is used to demonstrate this. If one cause is sufficient, the lines are simply combined. The safety barriers are then shown in a diagram. The diagram should show the possibility of an accident if all the safety measures along an event-sequence fail.

There are different ways to proceed in constructing such a diagram. One way (Taylor, 1994) is to start with concentrations of energy (hazard sources). The safety barriers surrounding each hazard source are listed. Also the intended combinations of safety barriers for each operational state are listed. In the analysis, the reliability of a barrier and the possibilities for bypassing it are investigated. The analysis also includes a check that criteria for "good" barriers are fulfilled. Such criteria are also given in the references to the method.

The approach was originally devised with chemical installations in mind, but it appears to have a wider area of application. It has several similarities with Fault Tree Analysis. One advantage is that a Safety Barrier Diagram probably is easier to understand by non-specialists.

A possible difficulty is that there are often a large number of potentially hazardous disturbances in any one system. This would involve the construction of a large number of diagrams unless the number of disturbances to analyse could be reduced through suitable grouping.

10.5 CONCEPT OF SAFETY FUNCTION

Varying terminology

The terminology used to describe safety features varies considerably. A number of terms with somewhat divergent meanings have been discussed. Some examples are given below, of which the four first appear in Section 10.2 and the other two in Section 11.6. They include:

- Barriers.
- Defences (Reason, 1997).
- Functional safety, i.e. the ability of a safety-related system to carry out the actions necessary to achieve a safe state for the equipment under control (IEC, 1998).
- Protection layer (defined by example) (CCPS, 1993).
- Barrier function, which can arrest accident/incident evolution so that the next event in the chain will not happen (Svenson, 1991, 2000).
- Barrier function systems, which are the systems performing the barrier functions (Svenson, 1991, 2000).

"Safety function" is a rather common term, but no clear definitions have been found in the literature. Even in the "Standard on Functional Safety" (IEC, 1998), where the term is used several times, it is not defined. It might therefore be used in different senses in various applications.

Proposal for a definition of safety function

A general definition of safety function is therefore proposed:

> *A safety function is a technical, organisational or combined function that can reduce the probability and/or consequences of accidents and other unwanted events in a system.*

Deliberately, "safety function" is defined as a broad concept. In principle it covers all the definitions and concepts presented earlier in this chapter. In specific applications it requires more concrete characterisation (as described below).

Figure 10.2 visualises the model and its basic components. The model might represent a company subject to a number of different hazards. These can cause different kinds of injury and damage. In order to prevent these, there is a set of safety functions. Hazards include energies, and different kinds of disturbances and deviations.

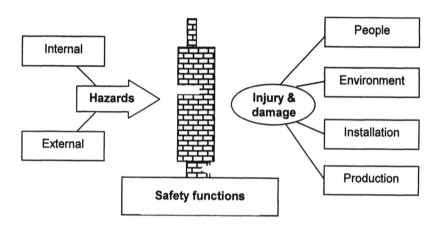

Figure 10.2   General model of safety functions.

Safety function parameters

For practical and operational applications any safety function (SF) can be described by a set of parameters. The most essential are:

a)   Level of abstraction.
b)   Systems level.
c)   Type of safety function.
d)   Type of object.

a) *Level of abstraction* starts at the lowest level with a concrete solution, e.g. a safety relay or a temperature guard. At a higher level, this can mean protection again excess temperature (functional solution). At a still higher level, it might involve process control (see also example in Section 10.7).

b) *Systems level* is related to the systems hierarchy. Examples of levels are component, subsystem, machine, department and a whole factory. The categorisation can be applied to the system under study, as well as to safety functions. The concept can also be extended to societal level so as to include laws regulating safety, fire brigades, emergency services in general, and so on.

c) *Type of safety function* describes what is included in a safety function. It can be divided into technical, organisational and human functions. Note that functions where safety is not the main objective may also have essential safety features. All these can be at different levels of a) and b).

d) *Type of object* characterises the object, i.e. the system that is to be safe. This may be a technical system, software, control room and related equipment, etc. Organisational conditions of different kinds should be included here. Examples include the management of projects and maintenance.

Characteristics of safety functions

A safety function (SF) can be described by a set of characteristics intended to describe its contribution to overall safety, and also provide a basis for its evaluation. Examples of relevant characterisations are:

- Consequences of the failure of a SF, which can describe how important the SF is, and also whether a failure leads directly to an accident, to a latent failure, or to something else.
- Robustness of the SF to deviations, interruption of procedures etc. (or the opposite—vulnerability, which might be more easy to handle).
- Opportunity or not to verify whether results of a SF agree with expected outcome.
- "Efficiency" is intended to give a measure of how well a SF can fulfil its aim. A general definition has not yet been formulated, and might be rather tricky if it shall cover all types of SF.

Similar terms can be found in a standard issued by the IEC (1998):

- *Functional safety* is the ability of a safety-related system to carry out the actions necessary to achieve a safety state for the "Equipment Under Control".
- *Safety integrity* is the probability of a safety-related system satisfactorily performing the required safety-related safety functions under all the stated conditions within a stated period of time.

Comments

The intention of the safety-function concept is to obtain a fairly simple framework as a basis for safety discussions and system analysis. Depending on the system concerned, a suitable set of characteristics can be defined. If the system is well known, and relations between different SFs can be modelled, probabilistic estimates of characteristics could be made.

Other sets of characteristics might be chosen. Threats and different factors that can reduce safety, e.g. active failures, latent conditions and local circumstances (Reason, 1997), might be included. Such aspects might be considered in the SF Model, but are not included in the characteristics listed above.

Practical application will indicate the best way to arrange the parameters, and also how detailed the descriptions should be. For example, an essential aspect consists in the characteristics of organisation (for parameters c and d). This can range from an organisational hierarchy with strict rules for decision-making, to informal and less clear ways to make agreements and decisions.

An example of a description of safety functions in a company is shown in the example in Section 10.7.

10.6 SAFETY FUNCTION ANALYSIS

The method

Based on the concept of safety function, a methodology called Safety Function Analysis has been developed (Harms-Ringdahl, 2000). The goals of a Safety Function Analysis (SFA) are to achieve:

- A structured description of a system's safety functions.
- An evaluation of their adequacy and weaknesses.
- Proposals for improvements, if required.

In principle, SFA has two general applications. The first takes the system (workplace) and its hazards as a starting point. The second is primarily an accident investigation, and is designed to draw conclusions about SFs and their weaknesses on the basis of an accident or near-accident event. This chapter generally addresses the first application.

SFA procedure

A SFA contains six main stages, and—like most safety analysis methods—also includes making preparations and concluding the analysis (Figure 10.3).

PREPARE
Before an analysis can be conducted, its aim and basic conditions need to be defined. This concerns for example:

- The boundaries of the system under study. What shall be included in the analysis?
- Identification of hazards. Before the analysis starts, fairly good knowledge of existing hazards is needed. Sometimes, it can be essential to conduct a more or less simple preliminary safety analysis, e.g. Energy Analysis.

- For which types of hazards shall the safety functions be studied? The analysis could be restricted to occupational accidents, but it could be enlarged to include, say, environmental damage.
- The operational conditions that are supposed to prevail.
- Performance of the analysis, in terms of available time and resources.

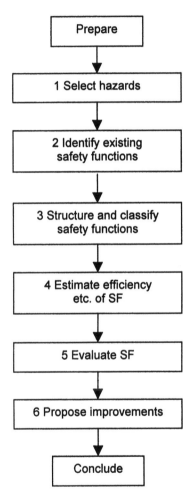

Figure 10.3 Main stages of procedure in Safety Function Analysis.

1. SELECT HAZARDS

A selection is made of the hazards for which safety functions are to be analysed. This can concern energies, essential deviations or disturbances. The amount of these may be large, so restricting oneself to the most relevant ones is essential. Otherwise, the analysis can be too heavy.

Results may be available from earlier safety analyses. Otherwise, a quick energy analysis could be performed.

2. IDENTIFY EXISTING SAFETY FUNCTIONS

One of several different approaches could be adopted here. One would be to start with a structured checklist of general safety functions, and to identify the ones that are relevant. Another would be to start from specific hazards and pose questions of the following kind:

- How is the likelihood of an accident kept low?
- How are consequences kept to a low level?
- How is the damage reduced if an accident should occur?

This procedure will give items at a fairly concrete level. Answers can be obtained from an interview or in a group discussion—resulting in a list of safety functions.

3. STRUCTURE AND CLASSIFY SAFETY FUNCTIONS

The list from the preceding stage needs to be arranged in a logical way to facilitate the estimations and evaluations that follow. There is no unique solution; rather, arranging is an iterative process aimed at achieving a simple and logical presentation. Figure 10.4 shows the results of such structuring.

In structuring, it might be of help to consider the proposed parameters of safety functions (see Section 10.5). These are:

a) Level of abstraction.
b) Systems level.
c) Type of safety function.
d) Type of object.

4. ESTIMATE EFFICIENCY, ETC.

The aim of this stage is to obtain estimates of some characteristics of the safety functions. Again, there are a number of alternatives. If sufficient information is available, more or less traditional estimates of reliability can be made. A simple way is to make classifications of "*Importance*" and "*Efficiency*"; "*Intention*" may also be useful.

Importance

Importance from a safety point of view might be categorised into four types. Here, it should be assumed that the SF works as it should. The first type (0) means in practice that it could be removed without affecting the probability and potential consequences of an accident:

0. No influence on safety.
1. Small influence on safety.
2. Rather large influence on safety.
3. Large influence on safety, closely connected to accidents and size of consequences.

The estimate can be made quite advanced, e.g. by looking at how large a contribution the SF makes to overall reliability. However, it is probably advisable to make a preliminary judgement in a simple manner. For the essential SFs, a more exhaustive analysis could be made.

Efficiency
"Efficiency" for each SF is estimated. It might be expressed as a success rate, which can be seen as a combination of the reliability of the function, and the probability that it is in place. This rate ranges from 0% for a function that is estimated to fail under all circumstances to over 99.99% for a function that works well.

"Success rate" will vary in meaning according to type of SF, and might involve quite complicated considerations in some cases. In the case of a technical system, rather straight-forward reliability considerations could be made. But if the function is performed through an informal organisational routine, how an estimate is to be made is not so obvious.

Intention
The "intention" of an SF can be included in the estimation. It might be especially valuable in design situations, when it is often essential to define intentions in accordance with different solutions. "Intention" might, for example, be divided into the four categories below:

a) No intended SF, and no influence on safety.
b) No intended SF, but some influence on safety.
c) Intended SF, but also other main functional purpose.
d) SF largely intended to improve safety (probability or reduction of consequences).

One problem is that original intention might not be known, giving too uncertain estimates. One advantage of discussing "intention" is that it can provide a good support in estimating the "success rate" as it defines what is the desired result.

5. EVALUATE SAFETY FUNCTIONS
Basically, this stage concerns whether the SFs are good enough, and if their coverage is sufficient to control the hazards concerned. For each SF a judgement should be made, if it is acceptable or if improvement is

recommended. The scale presented in Table 4.3 for direct risk assessments might be used here.

It is more difficult to evaluate the overall safety system, in particular to see if there are gaps in the system with regard to safety functions. In complicated cases, there might be a need also to apply risk-oriented methods for safety analysis so as better to understand the situation.

The output of this stage can be approval of the safety system, or parts of it. It can also be a recommendation to improve a certain SF and/or supplement it with one or others.

6. PROPOSE IMPROVEMENTS

Some SFs might need to be improved, with regard either to "efficiency" and/or the elimination of weak points. Improvements can also relate to the coverage of an SF with too narrow an application area.

CONCLUDE

The analysis is concluded by making a report. This summarises the analysis, and gives results, assumptions, and the basis for assessments.

Comments of SFA

The method is generic, which involves that final results can have quite different appearances. Dependence on approach and the analyst's skill is greater than for more traditional methods. For example, estimations can be made and presented in various ways. In practical applications it is essential to present the foundation for any estimation made. Then the reader of the analysis report has an opportunity to judge the rationality of the results obtained.

One way of making estimates (Stage 4) is to base them on interviews with people in different positions in the organisation. Their different perspectives have proved to give valuable input into the analysis. At the estimation and assessment stages, one option is to include "Don't know" as a response option (especially in the case of interviews).

In SFA both technical and organisational features are usually essential, and should be included in systems description.

Alternative applications

Other ways of performing a SFA can be chosen. One option is to base the analysis on an accident investigation—both for identification of SFs and for estimation of which SFs worked or did not work at the time of the accident.

Another alternative is to start with the modelling stages (2 and 3), but then continue adopting a quantitative approach. Stricter modelling would be demanded, including consideration of connections between different SFs. At the estimating stage, numerical values for efficiency and reliability would then be searched for and employed.

10.7 EXAMPLE OF SAFETY FUNCTION ANALYSIS

Background

In this example an existing production system is analysed. An earlier safety analysis revealed a number of hazards, which indicated the need for a number of corrections.

A similar system will be designed in the near future. The aim of the analysis is to obtain information that can support the design and planning of the new workplace. The intention is to make design and planning correct from the very beginning.

The studied system

The technical part of the production system consists of five similar production tanks, each with a volume of about 3 m^3. These are used to mix various compounds, and no chemical reactions should occur. The site also accommodates a cleaning system using lye and hot water, which are governed by a computer-control system.

In principle, simple batch production is involved, where different substances are added and mixed following strict procedures. Hygienic demands are high, and cleaning follows specified routines. An essential part of the work is manual, guided by formal procedures and batch protocols. In the workplace 20 people are employed in total, and production is run in shifts.

The workplace forms part of a large factory with an over-arching organisational hierarchy. This means that overall production planning also sets guidelines for health and safety work.

The analysis

PREPARE
The aim of the analysis is to obtain information that can support the design and planning of a new workplace of the same kind.

The whole workplace and its place in the company hierarchy form the object of the analysis. Both technical equipment and organisational aspects shall be included, but not down to a very detailed level.

1. SELECT HAZARDS

For the analysis, three hazards were selected. Two of these were lye (pH 13.5), and hot water (80°C) that could cause serious burn injuries. The third was the mechanical movements of a mixing screw that might cause crush injuries.

From earlier Energy Analysis and Deviation Analysis, a number of different possibilities for accidents to occur were already known.

2. IDENTIFY EXISTING SAFETY FUNCTIONS

Information about safety functions was collected in a dialogue with an engineer familiar with the system and its design history. He had also participated in earlier safety analysis.

Identification of the existing SFs was based on a discussion of a few accident scenarios. The first was the collapsing of the tank due to over-pressure. SFs that might prevent such an accident were identified. This was followed by a search for functions related to mitigation and emergency activities.

Supplementary identification came from a check against the parameters of the general model. This led to identification of some additional SFs not noted during the first round. Ultimately, a list of about 50 SFs was obtained.

3. STRUCTURE AND CLASSIFY SAFETY FUNCTIONS

The identified SFs were structured in six general groups, as shown in Figure 10.4. The SFs were entered into the record sheet (see Table 10.2), arranged in accordance with the obtained structure at a rather detailed level.

4. ESTIMATE EFFICIENCY, ETC.

At the estimation stage, the figure and the table were used to explain the meaning of SF and its manifestation in this workplace. For each SF, estimation was made of "Importance", based on the scale presented in Section 10.6:

0. No influence on safety.
1. Small influence.
2. Rather large influence.
3. Large influence, closely connected to accidents or size of consequence.

A further estimation was made of "Efficiency" (success rate), and numerical values between 0 and 1 were assigned.

During the estimation discussions, it was found that some extra SFs needed to be added to the list so as better to describe the system.

5. EVALUATE SAFETY FUNCTIONS

The analysis team evaluated each SF on the list. The judgement concerned whether the SF was acceptable, or if improvements were needed. The scale in Table 4.3 for direct risk assessments was used.

The coverage and scope of the safety system in general were regarded as insufficient to control the considered hazards. The conclusion drawn was that improved functions were needed, at both a detailed and general level.

6. PROPOSE IMPROVEMENTS
Proposals were made for the SFs that were not approved. In several cases, direct concrete solutions were found. However, for a number of SFs information was not sufficient, and the "proposal" simply consisted of a request for a further check. In particular, this concerned the computer control system, where the design was not transparent enough to allow any adequate proposal.

CONCLUDE
The analysis was summarised in a report, which described the results, the recommendations, the assumptions, and the basis for assessments.

The recommendations contained suggested improvements for the existing workplace since quite a lot of the SFs were found to be inadequate. The other part of the report applied to the new (planned) workplace, thereby increasing opportunities to obtain a correct design from the outset.

Model of safety functions

One essential part of results is the description of the safety functions relevant to the workplace. Such a description can be prepared in several ways. Here, it was based on the four parameters described in Section 10.4, namely:

a) Level of abstraction.
b) Systems level.
c) Type of safety function.
d) Type of object.

Illustration of four parameters might require a four-dimensional model, which is rather impractical. The four parameters were therefore reduced to two dimensions, as described in Figure 10.3.

The safety functions are divided into four levels—the columns—which combine level of abstraction and systems level.

1. The highest level is called *general function* and is related to the aim of the SF.
2. *Principal function* is a less abstract description.
3. *Functional solution* describes the functions in greater detail, and is at a lower systems level.

4. The *concrete solutions*, e.g. a specific safety relay or an operator's action, are at a lower systems level. These are not shown in the figure, but were listed on the record sheet (Table 10.2).

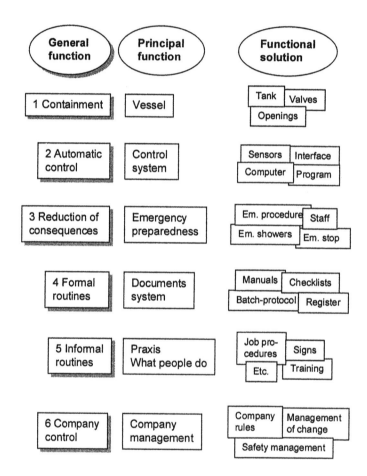

Figure 10.3 Model of safety functions in the workplace.

Structuring into six parts was based on the parameter c: *Type of safety function*. The major groups were as follows:

1. *Containment* refers to mechanical technical devices that separate the hazards (hot water, lye, and mechanical movements) from operators during normal operation.

2. *Automatic control* starts and stops movements, and contains for example a number of interlocks ensuring that openings in the tank are closed, etc.

Table 10.2 Extract from the record sheet of the Safety Function Analysis in the workplace.

Principal function / Functional solution	Concrete solution	Assessments*			Proposed measures
		Imp	Eff	Acc	
3 Emergency		-	-	3	
3.1 Emergency stop-button	Emergency-stop buttons	3	0.2	3	Include pumps & Emergency Package
3.2 Emergency showers	a) Shower (1 in the workplace)	3	0.1	3	Emergency Package
	b) Eye showers (3)	3	0	3	Emergency Package
3.3 Emergency procedure	Persons with special training	2	0.2	3	One person on all shifts
5 Informal routines		-	-	3	
5.1 Written instructions	a) Information	1	0.5	2	Improve, investigate needs
	b) Instructions for special activities	2	0.5	2	Include safety aspects
	c) Instructions for machines	2	0.3	3	Check and improve
	d) Instructions for control system	3	0.1	3	Rewrite completely, make user friendly
5.2 Oral instructions	a) "If disturbance, ask for help"	2	0.5	3	Include in 5.1.a
	b) "If cleaning in operation, do not enter"	3	0.5	3	Include in 5.1.a
5.3 Information signs	"Cleaning in operation"	2	0.5	3	Investigate to find better system
5.3 Training	a) General	1	-	1	
	b) Disturbances in control system	2	0.1	3	Include in improved instructions 5.1.d

* **Assessments:** Importance, Efficiency, and Acceptability (as above).

3. *Reduction of consequences* refers to a number of technical devices, e.g. for the prevention of overpressure, and organisational activities such as availability of emergency showers.

4. *Formal routines* are regulated in a system of documents, which are carefully worked out, formally approved, and strictly followed. The item here is related to production in the specific workplace under study.

5. *Informal routines* indicate features of the organisational system and what people do in practice in the actual workplace. The subject here is very wide and could be treated in several ways. Its content ranges from what operators do in their daily work through to verbal and written instructions (but not in the sense of formal routines).

6. *Company control* designates how safety instructions and rules emanate from the top of the company. For example, it includes the system for safety management (operated by this company).

Comments on results

Parts of the record sheet are shown in Table 10.2. The analysis revealed several weak points to the safety functions (both technical and organisational).

They concerned, for example, the emergency systems, which are partly shown on the record sheet. There were emergency showers, both for eyes and for the whole body. It was found that they would not be useful when needed—efficiency was scored low on both. They were too far away, and people were not aware that they had to use them quickly.

The proposed "Emergency package" included extra showers (for both body and eyes). An investigation was needed into where these should be located. Better information and training were also essential.

Some examples of organisational SFs classified as "Praxis, informal" are also shown in Table 10.2. Again, a number of improvements were needed.

In the analysis a number of important SFs were identified, which actually had purposes other than the support of safety. One was a hole in the tank, intended for input of material, which also provided ventilation for the tank and prevented overpressure. But when the hole was changed, it was not observed that the risk of overpressure increased substantially.

Several functions in the computer control programme were also essential to safety without being classified as such. Programme changes in this case might unintentionally decrease certain SFs.

This case is also discussed as an example of safety analysis in Section 15.8, where further information about results and analytic procedure are given.

11
Some further methods

11.1 INTRODUCTION

Nowadays, there is a wide range of methods for analysing system risks and safety properties on offer. The previous chapters have contained descriptions of selected methods for safety analysis, but they represent only a small number of all the conceivable methods available. Further, they represent the choice of the author. Others might well have made a different selection.

The aim of this chapter is to broaden the picture and provide an overview of further methods. They have been chosen so that many different ideas are represented, and so that descriptions available in English. Rather brief accounts are given, but references are also provided to enable the interested reader to go further.

The methods are arranged into six categories. However, there is an overlap between areas and some methods belong to more than one group.

a)   Technically oriented methods.
b)   Human oriented methods.
c)   Task analysis.
d)   Management oriented methods.
e)   Accident investigations.
f)   Coarse analyses.

The group Coarse analyses comes in the last, but there is a sense in which it is the most important category. The reason for this is that "rough" methods are used much more commonly than other types. Consequently, they deserve quite a lot of attention.

11.2 TECHNICALLY ORIENTED METHODS

There are several methods with a technical orientation. Energy Analysis, HAZOP, and Fault Tree Analysis belong to this category. This section gives further examples of such methods. However, some of them have a wider application, especially Event Tree Analysis. The methods discussed here are:

- Failure Mode and Effects Analysis (FMEA).
- Event Tree Analysis.
- Cause-Consequence Diagram.
- Reaction Matrix.
- Consequence analysis models.

Failure Mode and Effects Analysis (FMEA)

FMEA is a well established method, which has been utilised since the end of the 1950s. The method is well documented, and several descriptions of its use are available (e.g. Hammer, 1972; Taylor, 1994). There is also an international standard (IEC, 1985) available. This, however, has the character of a users' manual, providing guidance rather than being an attempt to standardise applications.

The method is employed for analyses of technical systems. In principle, every component in the system is examined, and two basic questions are asked:

- How can the unit fail?
- What happens then?

FMEA can be used at different system levels—from individual components to larger function blocks. This means that details of the analytical procedure will vary. The main stages in an analysis are as follows:

- The system is divided up into different units in the form of a block diagram or a list.
- Failure modes are identified for the various units.
- Conceivable causes, consequences and the significance of failure are assessed for each failure mode.
- An investigation is made into how the failure can be detected.
- Recommendations for suitable control measures are made.

It is best to use a special record sheet for the analysis. The IEC (1985) provides a version with 12 columns, where table headings include:

- Identification—component designation, function.
- Failure mode.
- Failure cause.
- Failure effect.
- Failure detection.
- Possible action.
- Probability and/or criticality level.

When using FMEA a large number of possible failures will be discovered. It is practical to make a classification of their importance, which for this method is often called "criticality". This can be accomplished in various ways, such as by weighing the probability of occurrence and the seriousness of effects. Sometimes, the method is called Failure Mode, Effects and Criticality Analysis (FMECA) which makes the criticality assessment explicit.

The IEC (1985) provides an example of a criticality scale, where the most serious level is: "Any event which could potentially cause the loss of primary system function(s) resulting in significant damage to the system or its environment, and/or cause the loss of life or limb."

A detailed analysis may be extensive. A system can contain a large number of components, and a component can fail in many different ways. A relay, for example, may have 15 different failure modes (Taylor, 1994). In the standard description (IEC, 1985) 33 generic failure modes are listed.

Event Tree Analysis

An event tree starts with an initiating event, e.g. a pump that ceases to operate, and then describes the consequences of this. Short descriptions are available (e.g. CCPS, 1985; Suokas and Rouhiainen, 1993; Lees, 1996). Often the trees are technically oriented, but an event tree can also include human actions.

The general procedure for Event Tree Analysis includes four steps:

1. Select an initiating event.
2. Identify the safety functions designed to deal with the initiating event.
3. Construct the event tree.
4. Describe the accident sequence (and eventually calculate accident frequency).

The principle can most easily be explained by example. Figure 11.1 shows an event tree where a dust explosion is the starting point. This might lead to a fire, and the tree shows alternative outcomes given two safety functions—a sprinkler system and a fire alarm. Frequency of the events H1 to H5 can be calculated if estimates are available for frequency of initiating explosion and reliability of sprinkler and alarm.

Event trees can be designed differently. Often the design starts from the left giving a lying tree, but Figure 11.2 starts from the top (appearing as a standing tree). What they have in common is that they, more or less strictly, show how events evolve over time.

Another example takes a container for toxic gas and a person working in a control room nearby. Since the container might leak, a gas detector has been installed. In the case of a leak, an alarm bell should sound prompting the person to rush out of the premises.

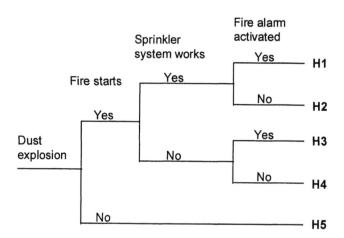

Figure 11.1. Event tree showing different routes in an accident sequence starting with a dust explosion (adapted from Rouhiainen, 1993).

Figure 11.2 illustrates an event tree for a sequence starting with a gas leak. Every part of this sequence contains the possibility of success or failure. All leaks will not necessarily lead to gas being present in the workplace—the first branch-off point. Failure can be due to the gas not reaching the detector. The alarm may fail to go off, or the person might not manage to get out. In this tree there are two possible end consequences: injury or no injury.

An event tree provides opportunities for making quantitative estimates. The initial event is expressed as a frequency (events per year). The branch-off points are expressed as probabilities (the number of failures per trial or occasion of use). Figure 11.2 provides an example of how an estimate can be made. For purposes of clarification, rather high frequency and failure probability values have been used. On the basis of these values, the frequency of damage to a person resulting from a gas leak is around 0.2 times per year $(0.1 + 0.04 + 0.072)$.

Cause-Consequence Diagrams

One technique, related to both Event Tree and Fault Tree Analysis, involves the use of Cause-Consequence Diagrams. The analysis starts with the definition of a critical event in the system. Possible causes of the event are then filled in, as with the construction of a fault tree. Consequences are also investigated, as when an event tree is constructed. Descriptions of the method have been provided by Nielsen (1971, 1974) and Taylor (1974, 1994).

One feature of the method is that alternative consequence paths can be handled. It can also be used as a basis for making probabilistic estimates.

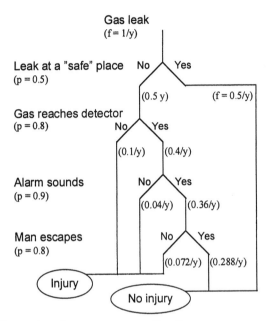

Figure 11.2 Example of an event tree for the consequences of a gas leak (f = frequency, y = year).

Reaction Matrix Analysis

In a workplace where several chemicals are present, there is a danger of unwanted chemical reactions. Reaction Matrix Analysis is a technique designed to aid identification of dangerous combinations. There are brief descriptions of the method available (e.g. Suokas and Rouhiainen, 1993; Taylor, 1994).

The basic idea is to construct a matrix with rows and columns labelled according to the chemical substances in the workplace. (The rows and columns will have the same labels.) For each combination of chemicals reaction potential is estimated and entered into the matrix. In its simplest form, an X can be entered to show a potential danger. More detailed information can also be given in each matrix cell.

In order to perform a matrix analysis, good knowledge of chemistry is needed. Some limitations of the method are that the chemical materials are assumed to be pure, and that the method itself only considers combinations of two chemicals.

Consequence analysis models

Especially with regard to accidents involving chemicals, analysis of consequences is of great importance. An analysis might concern:

- A fire.
- An explosion.
- The release of toxic gases.
- The determination of toxic effects.

Often such phenomena are complex and require advanced considerations and calculations. The methodology goes beyond the scope of this book, but there is a large literature in the area (for overviews, see Taylor, 1994; Lees, 1996). Several computer programmes for calculations of different accident scenarios are available. However, there may be problems if the user has insufficient knowledge of physics or chemistry and the underlying analytic assumptions. The results might look tidy and professional, but may still be wrong.

11.3 HUMAN-ORIENTED METHODS

Human error methods

A general discussion of human errors was presented in Section 2.2 above. There are a number of techniques available for analysing human errors and tasks. Many of these are advanced and quite difficult to apply. The general field of human error analysis has become a specialist's area with a large literature. Only a short orientation is provided here.

In general, the aim of any such analysis is to predict human errors in a defined task, and consider what can go wrong. An analysis might, for example, study some specified operations in a control room, or how a specific problem is solved.

This section presents brief accounts of some methods in this area:

- Human Reliability Assessment.
- Human Error Identification (Action Error Method).
- Technique for Human Error Rate Prediction (THERP)—an example of a quantitative method.
- Cognitive Reliability and Error Analysis Method (CREAM).
- HAZOP—as extended to include human errors.

A short summary of a fairly general methodology called Task Analysis is provided separately (Section 11.4). One of its applications is to provide input into a form of human error analysis.

Human Reliability Assessment

One special area is concerned with probabilistic aspects of human errors, and is usually referred to as Human Reliability Assessment (HRA). It involves reliability engineers and human-factors specialists, and is applied mainly in the nuclear power domain (for overviews, see Kirwan, 1994; Gertman and Blackman, 1994). THERP, as described later in this section, is an example of an HRA method.

The focus is usually on quantification, and results are used in probabilistic safety assessments. The objective of HRA is to find the probability that an activity is successfully completed (or that it fails).

Kirwan (1994) has described the HRA process in terms of eight principal components:

1. Problem definition.
2. Task analysis.
3. Human error identification.
4. Representation of this information in a form which allows quantitative evaluation of the error's impact on the system.
5. Human error quantification.
6. Impact assessment, calculation of the overall system risk level.
7. Error reduction analysis.
8. Documentation and quality assurance.

However, there are a great number of human reliability methods with different procedures. Hollnagel (1993), for example, has published a list of 27 different HRA methods.

Identification of human errors

There are a number of similar methods aiming at the identification of human errors (e.g. Embrey, 1994; Kirwan, 1994). The methods are best suited for use in installations where there are well-defined procedures, e.g. in certain processing industries. If there are no well-established routines, it is difficult to find a basis on which an analysis can be conducted. In general, the aim is to identify steps that are especially susceptible to human error and assess the consequences of such errors.

One example is the Action Error Method described by Taylor (1979). The stages of analysis are:

1. Making a list of the steps in the operational procedure. The list specifies the effects of different actions on the installation. It must be detailed, containing items such as "Press Button A" or "Turn Valve B".

2. Identification of possible errors for each step, using a checklist of errors (see below).
3. Assessment of the consequences of the errors.
4. Investigation of conceivable causes of important errors.
5. Analysis of possible actions designed to gain control over the process.

Various conceivable types of errors include:

1. Actions not taken.
2. Actions taken in the wrong order.
3. Erroneous actions.
4. Actions applied to the wrong object.
5. Actions taken too late or too early.
6. Too many or too few actions taken.
7. Actions with an effect in the wrong direction.
8. Actions with an effect of the wrong magnitude.
9. Decision failures in relation to actions taken.

Taylor (1994) later developed a more detailed version of the Action Error Method. There are also several other methods for which similar approaches are adopted.

Technique for Human Error Rate Prediction (THERP)

THERP is a method for analysing and quantifying probabilities of human error, and is mainly used in the nuclear field. There is a handbook in which the method is extensively described (Swain and Guttman, 1983), and descriptions are also contained in other publications (e.g. Bell and Swain, 1983). The method has been developed steadily over a number of years. The main stages of the technique are:

1. Identification of system functions that are sensitive to human error.
2. Analysis of the job tasks that relate to the sensitive functions.
3. Estimation of error probabilities.
4. Estimation of the effects of human errors.
5. When applied at the design stage, utilisation of the results for system changes. These changes then need to be assessed further.

The handbook also contains tables with estimates of error probabilities for different types of errors. These probabilities may be affected by so-called "Performance Shaping Factors", meaning that the analyst makes adjustments to their values in the light of the quality of the man-machine interface, experience of the individual operator, etc.

Cognitive Reliability and Error Analysis Method (CREAM)

CREAM is based on a model and classification schema that can be used for accident investigation and the prediction of human performance. One essential element in the model is that a person tries to maintain control of a situation. The actions taken are determined more by the actual situation than by internal psychological mechanisms of failure. A distinction has been made between four control modes:

- *Scrambled control*; the choice of next action is in practice unpredictable and random.
- *Opportunistic control*; next action is determined by the current context rather than by stable intentions.
- *Tactical control*; this is based on planning, and follows a more or less known procedure.
- *Strategic control*; global context and higher level goals are considered.

An extensive description of the theory and method has been published (Hollnagel, 1998). In brief, CREAM is based on the following principles:

- The probability of human error depends on situation and context. Human errors cannot be analysed as isolated events.
- The probability that an error leads to an accident depends on the functions and state of the system.
- Prediction of future accidents and errors should be based on analysis and understanding of earlier incidents. A similar methodology is needed for near-accident investigation and predictive analysis.

Extended HAZOP approach

The principles of the method HAZOP (Chapter 8) are attractive for application to human errors, and there are some examples of different ways of proceeding (e.g. Kirwan, 1994). One way is to examine a process involving human actions, and apply the HAZOP guide words to that. An alternative is to apply HAZOP to a technical object, but also include human errors.

One example is the proposal made by Schurman and Fleger (1994) to incorporate human error into a standard HAZOP study. The analytic procedure is similar to a pure technical application, and human factor aspects are simply added. The major change lies in the incorporation of human-factor guide words and parameters.

The guide words are additions and reformulations of the standard HAZOP set. They include *Missing*, *Skipped* and *Mistimed*. The new/revised parameters include *Person*, *Action*, *Procedure*, etc. By combining guide words and parameters, meaningful and essential deviations can be detected. Schurman and Fleger state that the major adjustment needed is in the thinking of the analysis team. Operators and maintenance people should be regarded as subsystems of the process.

Deviation Analysis

Human errors have also been included in Deviation Analysis (described in Chapter 7). The approach is to treat human errors at the same time and in a similar manner as technical faults. This means that human actions are studied in less detail than in the more specialised methods. As support for the analysis team, there is a list comprising seven different categories of errors.

Even though this approach is quite simple, it offers a way of including human errors in an analysis in a practical and fairly simple manner.

Comments

A general problem with this kind of analysis is that the number of potential human errors can be immense, especially if multiple errors and advanced faults (e.g. in problem solving, etc.) are included. Ways of prioritising and limiting the number of potential errors become essential. Usually the human tasks to be analysed need to be precisely defined in any practical type of analysis.

The analysis of human errors is highly complex, and becomes even more complicated if calculating the probability that actions will go wrong is envisaged. A number of doubts have arisen concerning such calculations. Hollnagel (1993; 2000) points to the assumptions that have to be made, e.g. that it is meaningful to consider actions one by one, and that it is possible to determine a basic probability for a characteristic type of action. There is a question mark over how well such assumptions accord with reality. A general summary of these aspects is that human performance cannot be understood by decomposing it into parts, but only by considering it as a whole embedded in a meaningful context (Hollnagel, 1993).

11.4 TASK ANALYSIS

Task analysis is a methodology that covers a variety of human factors techniques. A large number of methods exist, and only a brief overview is given here. There are a number of fairly extensive reviews (e.g. Kirwan and Ainsworth, 1993; Annet and Stanton, 2000), and also more condensed summaries (e.g. Embrey, 1994).

The methods are aimed at what an individual does, especially manual workers and process operators, and sometimes also a team of operators. A division can be made into:

- *Action oriented approaches*, which give descriptions of the operator's behaviour at different levels of detail, together with indications of the structure of the task.
- *Cognitive approaches,* which focus on the mental processes that underlie observable behaviour and might include decision-making and problem-solving.

Hierarchical Task Analysis

Hierarchical Task Analysis (HTA) is a generic method for analysing how work is organised (Annet *et al.*, 1971). Basically, HTA involves the identification of the overall goal of the task, and then the various subtasks arranged in a hierarchy of operations. Results can be presented as a diagram, or in tabular format.

HTA starts by stating the objective a person should achieve. This is then broken down into a set of suboperations and a plan specifying when they are carried out. Each suboperation can be further divided if this is regarded as essential. Figure 11.3 represents the example of a computer-controlled lathe (see Figure 7.3) as a HTA.

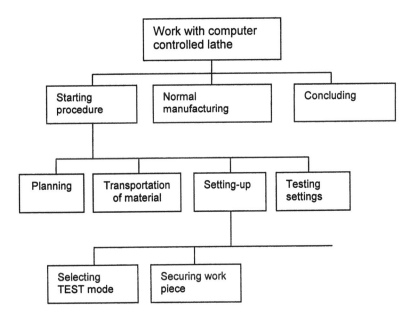

Figure 11.3 Hierarchical Task Analysis of work at a computer-controlled lathe (based on Figure 7.3).

Constraints associated with goals and task elements are analysed, which might influence the outcome of the task. If the task is critical, potential problems might be reduced by re-design, training, and so on.

HTA is probably the most common method in the group of action-oriented techniques. There are also a number of related methods (see Embrey, 1994). These include:

- Operator Action Event Tree, a special case of Event Tree (described in Section 11.2).
- Decision/Action Flow Diagram.
- Operational Sequence Diagram.
- Signal-Flow Graph Analysis.

Cognitive Task Analysis

Cognitive Task Analysis (CTA) methods address the underlying mental processes that give rise to errors. This can be essential, especially with regard to higher level functions (such as diagnosis and problem solving). The application of methodology is much more problematic here, since causes of cognitive errors are less well understood than action errors.

Applications of Task Analysis

Task analysis has several purposes. For example, it might be used for improving design of operational procedures in a control room. In the construction of human-computer interfaces, it is used to give a better under-standing of user demands and how different tasks should be allocated.

Another application is for the planning of education/training programmes in order to obtain good knowledge of the actions involved, e.g. in problem solving.

Task Analysis in itself is not designed to find risks, but it might provide an input into other safety analysis methods. The structured description of tasks can be useful in human error analysis in general, and it also fits well with Deviation Analysis.

Comments

Task analysis can offer valuable support in assessing and controlling risks. It is in a way a specialist's tool, and for several applications a fairly deep understanding of methodology is needed. It has been argued that there is a craft-skill requirement for conducting a sensitive HTA, arising from subjectivity in interpretation of data and ambiguity in the analytical process (Annet and Stanton, 2000).

There are a large number of methods, which may sometimes confuse potential users. Annet and Stanton (2000), for example, identify more than 100 task-analysis related methods. There is a concern within ergonomics that many of the methods developed are only ever used by their developers and have little significance for others. The authors would like to see methodological convergence, leaving a core of methods that satisfy most needs. They also point out that little is known at present of the reliability and validity of these methods.

11.5 MANAGEMENT ORIENTED METHODS

General

Organisational activities govern how an installation is designed, how work is carried out, who works at the plant, what safety routines there are, and so on. The quality and focus of these activities have decisive importance for the existence of hazards and how risks are controlled.

For this reason, it is important to have methods available for the analysis and assessment of the safety work of organisations. At the same time, it is a difficult subject—for a variety of reasons. Organisations and activities are not tangible objects, and it is not easy to get a grip on them. Written documentation reveals only a part of the reality. What causes difficulty is that there are informal decision-making paths, people with varying views on what is relevant, etc.

This section takes up examples of methods for examining organisational characteristics of a company. The list includes:

- Audits—in general.
- Management Oversight and Risk Tree (MORT).
- International Safety Rating System (ISRS).
- Safety Health and Environment (SHE) audit.
- Safety Culture Hazard and Operability Study (SCHAZOP).

Audits—in general

Audit has become a generally used term, but it does have a variety of meanings. A strict one is related to checking policy and intention in a company against how it actually operates. Another is examination of the management system to see if it conforms to some kind of (external) norm.

A definition of audit is given in a standard related to occupational health and safety (OH&S) management systems (BSI, 1996):

Audit is a systematic, and wherever possible, independent examination to determine whether activities and related results conform to planned arrangements and whether these arrangements are implemented effectively and are suitable to achieve the organisation's policy and objectives. "Independent" here does not necessarily mean external to the organisation.

The standard gives some general advice. Compared with routine monitoring, an audit should enable a deeper and more critical appraisal of all elements in a health and safety management system. The approach should be tailored to the size of the organisation and its hazards. Four general questions should be covered:

- Is the organisation's overall OH&S management system capable of achieving the required standards of OH&S performance?
- Is the organisation fulfilling all its obligations with regard to OH&S?
- What are the strengths and weaknesses of the system?
- Is the organisation (or part of it) actually doing and achieving what it claims to do?

There are other examples of advice related to audits of safety and health systems (e.g. Health and Safety Executive, 1991). Also general standards can be valuable in this context, since they address general aspects, such as standards for quality systems (ISO, 1990).

MORT

An almost classical method is MORT, which means Management Oversight and Risk Tree. Development of the method dates from 1970. A detailed guide and an account of the reasons for using MORT have been prepared by Johnson (1980). There is also a rather more summary description available (Know and Eicher, 1976).

"MORT emphasises that when an accident reveals errors, it is the system which fails. People operating a system cannot do the things expected of them because directives and criteria are less than adequate. Error is defined as any significant deviation from a previously established or expected standard of human performance that results in unwanted delay, difficulty, problem, trouble, incident, accident, malfunction or failure" (Johnson, 1980).

The energy model is an important element in MORT, and the MORT logic diagram can be seen as a model of an ideal safety programme. It can be used for:

- The investigation of an accident.
- The analysis of an organisational programme for safety.

The MORT tree

The MORT logic diagram provides a general problem description. It is rather like a fault tree and the same symbols are used. A small part of a MORT tree is shown in Figure 11.4.

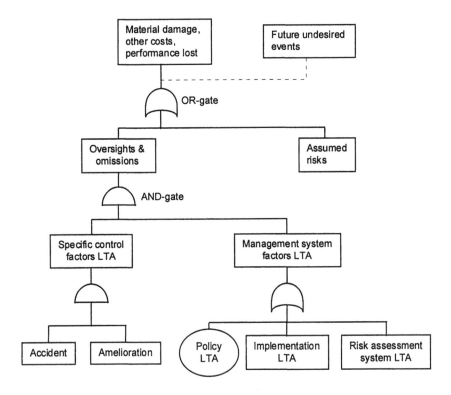

Figure 11.4 The top of the MORT tree (adapted from Johnson, 1980).

The top event may be an accident that has occurred. This can be due to an "assumed" hazard or to an "oversight or omission" (the two main branches of the tree), or both.

For a risk to be "assumed", it must have been analysed and treated as such by company management. Thus, the combination where a certain type of accident tends to occur and no specific control measure has been taken is not sufficient for the hazard to be counted as assumed.

The other main branch of the tree takes up organisational factors, and is called "Oversights and omissions". It has two subsidiary branches, one of which is called "Specific control factors" and focuses on what occurred during the accident. This is further divided into the accident itself and how its consequences are reduced, e.g. through fire fighting, provision of medical treatment, etc. The second subsidiary branch treats "Management system factors" and focuses on the question "Why?". It is divided into three further elements: policy, implementation, and risk assessment systems.

The various elements in the tree are numbered. These numbers refer to a list, which is provided as a complement to the tree. For each element there are specific questions that the analyst should pose. The tree contains around 200 basic problems. But, if it is applied in different areas, the number of potential causes it describes can rise to 1500.

Assessment (LTA)

Analysis involves going through the elements in the tree and making an assessment of each. There are two assessment levels: "Satisfactory" and "Less Than Adequate" (LTA). Assessments are in part subjective; i.e. different people may make different judgements. Nevertheless, the availability of a list of specific, and often concrete, questions for each element reduces the degree of subjectivity.

Procedure

The analysis is conducted by following the MORT chart, first in general and then in greater detail. Questions for which it is possible to find direct answers are marked. Colours are used to code the answers. Green means OK, red "LTA", and blue that no answer to the question has been obtained. Irrelevant questions are crossed out. The analysis is complete when all elements have been covered.

Comments

The methods permit a large number of problems to be identified. Johnson (1980) mentions that five MORT studies of serious accidents led to the identification of 197 problems, i.e. about 38 problems per study. He describes the method as simple; it is extensive, but each element is easy to understand. However, many perceive the method as impracticable, perhaps because there are so many different items to keep track of.

Johnson suggests that the analysis of an accident can be conducted in one or a few days. However, experiences from Finnish applications of MORT to maintenance work (Ruuhilehto, 1993) indicate that an analysis will require up to eight man-weeks.

MORT makes use of penetrating questions based on an ideal model of an organisation. Where the actual organisation deviates from the ideal, there can be far too many negative answers, which analysts lacking a great deal of experience may find difficult to handle.

International Safety Rating System (ISRS)

ISRS is a commercially available audit system, which has been widely used. A company can buy a licence to use the method, and this can be connected to consulting services. Accordingly, the full manual is not publicly available but there are shorter descriptions (e.g. DNV, 1990).

The objective of an ISRS is to obtain a measure of the effectiveness of a company's safety activities compared with a set of criteria developed for ISRS. A further aim is to provide a system to guide the development of an effective safety programme.

The ISRS audit consists of around 600 questions, which are divided between 20 elements. Each question is given a score for compliance with a given procedure or practice. Scoring guidelines are provided in the audit manual. Examples of the 20 elements include:

- Leadership and administration.
- Management training.
- Planned inspection.
- Task analysis and procedures.
- Accident/incident investigations.
- Planned task observation.
- Emergency preparedness.

Each element is structured into smaller parts. For example, "Leadership and administration" consists of 13 sections, each of which contain a number of questions:

1.1   General policy.
1.2   Programme co-ordinator.
1.3   Senior and Middle Management participation.
1.4   Established management performance standards.
        etc.

In an application, the company can choose which level of compliance it will aim at. There are ten possible levels; it could be Standard level 1 to 5 stars, or Advanced level with 1 to 5 stars.

Due to the widespread use of ISRS, some evaluations have been made (e.g. Eisner and Leger, 1988; Guastello, 1991; Chaplin and Hale, 1998). However, it is not easy to make a simple summary of these.

Safety, Health and Environment (SHE) audit

A guideline for auditing safety, health and the environment (SHE) can be given as a further example (Association of Swedish Chemical Industries, 1996). Its core consists of a performance guide with criteria related to SHE management. They are constructed with the "good praxis" management system as a model.

It is a "free" model and not directly based on any formal system for safety, health and environment, but it contains essentially all the points that are common to any environment system, e.g. EMAS (CEC, 1993) and ISO 14001 (ISO, 1996). The model does not presuppose such a system, and it can also be applied to less formally structured systems.

The performance guide covers around 140 activities, divided into General (90), Health (25) and Environment (30). Each of these can be evaluated, and then given a score ranging between 0 and 10.

The score is obtained by choosing the alternative that best agrees with the current situation. Interpolation between scores is sometimes necessary. An example:

Element 1.1.1 Policy

Score 2:   No written policy.

Score 4:   Written policy but of low quality and only limited knowledge of it in the organisation.

Score 7:   Policy of good quality in all areas (SHE). The policy is presented and explained to all employees and also to contractors.

Score 10:   The policy is well implemented and regularly used in training. Regular auditing. Policy is supported by local documents for interpretation and transfer into practical use.

The system can be used according to the aspirations of the company, and a subset can be used for application in smaller companies.

Safety Culture Hazard and Operability Study (SCHAZOP)

Ideas from the method HAZOP (Chapter 8) have also been applied to safety management systems. Kennedy and Kirwan (1998) have proposed a method called Safety Culture Hazard and Operability (SCHAZOP). The approach attempts to identify:

- Areas where the safety management process is "vulnerable" to failures.
- Potential consequences of failures.
- Potential (safety culture) "failure mechanisms".
- Factors influencing the likelihood of failures.

The SCHAZOP procedure includes three main stages:

1. Representation of the safety management process; this might be achieved by means of Hierarchical Task Analysis (see Section 11.4).

2. Selection of work group, plus guide words and property words.
3. The SCHAZOP study meeting.

The authors suggest eight guide words: Missing, Skipped, Mis-timed, More, Less, Wrong, As well as, and Other. They also propose a number of "property words", including Person/Skill, Action, Procedure/Specification.

Some case studies have been performed (Kennedy and Kirwan, 1998), and have been judged to function well. However, they were resource intensive and 300 hours of person effort were required for a study.

Deviation Analysis

Organisational aspects are included in Deviation Analysis, as described in Chapter 7. The aim is to identify essential deviations from company rules, and good praxis in planning and organisation. Compared with other methods, "normal" management activities are analysed and not only safety management.

There are two somewhat different approaches. One starts with potential technical or human deviations, and the analysis looks for management aspects that increase or decrease risks. The other examines organisational activities in order to identify deviations that are essential to safety. As support for the analysis team, there is a list comprising seven different categories of management activities.

The approach is quite elementary, but offers a way to include organisational activities in an analysis in a fairly simple manner.

Comments

Analysis of management is quite a complicated area. Many of the methods are based on practical experiences and ideas, which are organised in a structured way. Other methods depart from a more theoretical perspective.

The six methods described can be placed in three different groups.

1. Safety management is compared with a given norm, and agreement with the norm is assessed in Yes/No terms or by means of point-scores (MORT, ISRS, and SHE audit).
2. The safety management system at a company is described (modelled), and problems and deviations are then identified (General audit, and SCHAZOP).
3. The activity, job or other subject of analysis is described (modelled). In the analysis potential failures in the organisation are identified. All organisational functions are encompassed, including safety management (Deviation Analysis).

There are several further methods and concepts concerned with the analysis of safety management and safety culture, some of which were initially developed for the nuclear industry. The list is long, but some methods, with acronyms and references, are added here to give a more complete picture:

- ASCOT—Assessment of Safety Culture in Organisation Team (IAEA, 1994).
- CHASE—Complete Health and Safety Evaluation (Both *et al.*, 1987).
- Five Star System (British Safety Council, 1988).
- MANAGER MANagement Assessment Guidelines in the Evaluation of Risk (Pitbaldo et al., 1990).
- PRIMA—Process Risk Management Audit (Hurst *et al.*, 1996).
- SADT—Structured Analysis and Design Technique (Hale *et al.*, 1997).
- TRIPOD (acronym not explained) (Wagenaar *et al.*, 1994; Reason, 1997).

11.6 ACCIDENT INVESTIGATIONS

Introduction

An accident investigation can be seen as a safety analysis, given the broad definition applied here. A thorough investigation can provide useful information about the system in which an accident has occurred, and how to prevent further accident occurrences. The disadvantage from a methodological perspective is that the starting point for the investigation is a (more or less) random single event.

There are several methods for accident investigations, which are based on different principles. The choice of a suitable method and approach depends on the grounds for the investigation. Examples of aims and situations are:

a) Find out what happened with a quick and simple investigation.
b) Define responsibilities for the accident, which might concern regulatory aspects, financial compensation to injured people, and so on.
c) In the case of large accidents, obtain satisfactory understanding and explanation, and pursue a thorough and detailed investigation.
d) Conduct investigation as part of a plan to collect information about weaknesses in the system.
e) In systems intended to have a high level of safety, regard any accident as a systems failure; the investigation will then provide an opportunity further to improve the system.

Selection of approach will depend on the investigator's perspective on sources of accidents (compare Figure 2.2). Examples d) and e) are directed at obtaining information about the system, e.g. a workplace, and finding improvements. Both accidents and near-accidents can be studied.

There is a fairly large literature on accident and incident investigations, and how information can be used systematically to improve safety and prevent accidents (e.g. Schaff *et al.*, 1991; Kjellén, 2000). There are also a number of detailed guidelines (e.g. Hendrick and Benner, 1987; Ferry, 1988) of relevance to situations b) and c).

Methods

There are a number of methods that can be used for investigations. Some of the methods described in this book can be used for both system analysis and accident/incident investigations. These are:

- Deviation Analysis (Section 7.5).
- Event Tree Analysis (Section 11.2).
- Fault Tree Analysis (Chapter 9).
- MORT (Section 11.5).
- Safety Function Analysis (Section 10.6).

Some additional methods are presented in this section:

- Accident Evolution and Barrier Function (the AEB Model).
- Change Analysis.
- Multilinear Events Sequencing.
- STEP.

Accident Evolution and Barrier Function (AEB)

The AEB method can be used for analysis of accidents and incidents (near-accidents). AEB models an accident or incident as a series of interactions between human and technical systems (Svenson, 1991, 2000). An accident is described as a sequence of human and technical errors. In principle, there is the possibility of arresting development between any two successive errors.

The AEB method is related to safety barriers and functions, as discussed in Chapter 10. A central concept is "barrier function", which is a function that can interrupt accident/incident evolution so that the next event in the chain will not happen. A barrier function is always identified in relation to the system(s) it protects, has protected or could have protected.

"Barrier function systems" are the systems performing the barrier functions. A system might consist of an operator, an instruction, a physical separation, an emergency control system, or other safety-related systems.

The analysis is performed in eight steps according to the manual for the AEB method (Svenson, 2000). The result of an AEB analysis is a description of accident evolution as a flow diagram, which shows human and technical errors (Figure 11.5). A division is made in the "Human Factors System" and the "Technical System". The diagram also shows the barrier functions related to specific errors. If a particular accident/incident should occur, all the barrier functions in the sequence must have been broken or ineffective.

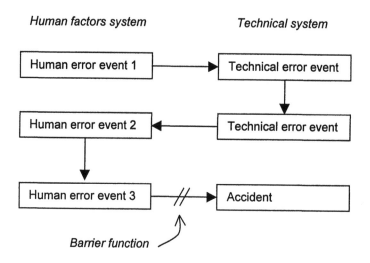

Figure 11.5 In the AEB-model, failures, malfunctions and errors are located as error events in boxes (adapted from Svenson, 2000).

An important purpose of an AEB analysis is to identify broken barrier functions and suggest how they can be improved. They are divided into three main categories:

- Ineffective barrier functions—ineffective in the sense that they did not prevent the development of an accident/incident.
- Non-existing barrier functions—if present they would have stopped the accident/incident evolution.
- Effective barrier functions, which actually prevented the progress towards an accident/incident. These are normally not included in an AEB analysis, since the AEB model is based on errors.

Change Analysis

Changes to a system can create new hazards or a deterioration in the control of hazards that are already handled. Change Analysis is designed to identify causes of increased risks arising from system changes. The method was originally designed for application to organisational systems (Kepner and Tregoe, 1965). Its aim is to identify the basic changes that give rise to problems. Change Analysis has been employed since the 1960s.

One application is in accident investigations (Bullock, 1976; Johnson, 1980; Ferry, 1988). The basic principle is to compare the situation on the occasion of the accident with a non-accident situation. The most important steps in an analysis are:

- Comparison of the accident and non-accident situations.
- Noting all known differences.
- Evaluation of the differences, and assessment of their influence on the accident sequence.

A special record sheet can be used for Change Analysis. It shows the factors that may be subject to change. For each factor, there should be descriptions of current and previous situations, differences, and changes that may have an effect. Twenty-five factors are divided into eight main groups:

1. What.
2. Where.
3. When.
4. Who.
5. Task.
6. Work Conditions.
7. Trigger Event.
8. Managerial Controls.

Both planned and unforeseen changes are included, and the method has similarities to Deviation Investigation. However, in Change Analysis it is assumed, more or less implicitly, that the old system has an adequate level of safety. This is sometimes a weakness of the analysis.

Multilinear Events Sequencing

A number of methods focus on the chronological sequence of an accident. There are several approaches, and what is looked for depends on the conceptual model underlying the analysis.

Multilinear Events Sequencing has been extensively described by Ferry (1988). It is based on the view that an incident begins when a stable situation is disturbed. A series of events can then lead to an accident. These can be plotted in a diagram where the actions of different actors are shown on a long a time axis. Figure 11.6 sketches out the principle for two actors. Conditions that influence the events can be inserted into the diagram. The logic chart can also be used to identify preventive counter-measures.

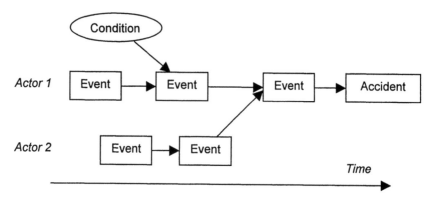

Figure 11.6. Activity events leading to accident for two actors (adapted from Ferry, 1988).

Sequentially Timed Events Plotting (STEP)

STEP is an event-based methodology for the investigation of accidents. An extensive manual, including advice on how to collect data, etc., describes the method (Hendrick and Benner, 1987). It is closely related to Multilinear Events Sequencing. An essential part of the procedure is to prepare a diagram of the sequence of events showing all the actors involved, which might also include relevant witnesses.

11.7 COARSE ANALYSES

Why perform a coarse analysis?

Even simple and quick analyses are of value and provide information on existing hazards. In many situations there is no justification for a stricter form of safety analysis, or the opportunities for one are lacking. Such situations can include:

- Presence of major safety deficiencies. If it is already known that there are a large number of safety problems, no detailed analysis is needed for these to be identified.
- Unclear picture of hazards. It is not known whether a thorough investigation is justified.

- Lack of resources. A full analysis cannot be conducted because of lack of people or time.
- Absence of documentation on the existing system or planned changes. There is insufficient information available for a proper analysis to be conducted.

A coarse analysis tends to have the following features:

- It is quicker to conduct than a normal safety analysis.
- It is less systematic, the methodology is often more free, and results are more difficult to repeat.
- It has limited coverage, meaning that only certain aspects of the system are considered, or that only specific types of hazards are investigated.
- It is usually intended to cover an entire system (which is an advantage).

Several of the methods already described can be used in a "coarse" or "quick" manner to cut down the time taken by an analysis. A short summary of a variety of approaches is provided below. Sometimes, the methods will overlap, so that one approach can contain some of the elements of another.

1. Use of checklists—based on summaries of known problems.
2. Inventories of documented hazards.
3. Inventories of known hazards.
4. Comparisons with similar installations.
5. Comparisons with directives and norms.
6. Preliminary hazard analysis.
7. "What-if."
8. Coarse Energy Analysis.
9. Coarse Deviation Analysis.

(1) Use of checklists

Checklists have been developed for a variety of situations and specific industrial sectors. The principle is to take every point on the list to see whether or not a particular hazard exists.

The quality and utility of any analysis depend very much on the checklist, and on whether the installation and its features are in agreement. If several similar installations are to be studied, a more thorough analysis can be performed on the first. The results from that are then used to make a special checklist for the remaining installations. Examples (2) and (3) below provide ways of obtaining suitable checklists

(2) Inventories of documented hazards

In the case of installations that have been in operation for some time, there is generally documentary material available on injuries and damage, e.g. accident investigation reports. An inventory can then be obtained by preparing a structured summary of these. The inventory can also be designed as a checklist, which might be used to examine whether hazards are properly handled.

(3) Inventories of known hazards

The purpose of taking an "inventory of known hazards" is to summarise the hazards that are known to employees and can be reported spontaneously. Such an inventory can be taken in many ways. One is to arrange a meeting of a suitably composed team, which then focuses on the most serious accident hazards in various parts of the installation. Good experiences have been had of "brain-storming". For example, one way of beginning is to have all participants write down the ten hazards they consider to be most serious. Again, results can be used for writing checklists.

(4) Checking against directives and norms

Directives issued by the authorities can sometimes be treated as checklists. They represent a summary of knowledge obtained over a long period of time. For example, a standard concerned with safety of machinery (CEN, 1996) provides a long checklist of hazards.

(5) Comparisons with similar installations

If there is a similar installation where hazards have been thoroughly investigated, this can be of good help. An analysis is performed of whether the same hazards exist at the object under study. This form of analysis may be appropriate when a new installation is planned, or when changes are made to an existing installation.

(6) Preliminary Hazard Analysis (PHA)

PHA has several different meanings. Sometimes, it signifies a rather large analysis, but usually it refers to a small analysis with a more or less defined approach. It may be wise to avoid the term, since it has several possible interpretations. Otherwise, a precise reference to the approach adopted can be made.

One early description was made by Hammer (1980), who introduced the term "Preliminary Hazard Analysis". The first step is to summarise known problems. The function of the system and its surrounding conditions are then examined, and an attempt is made to identify more serious hazards. When applied in full, the method is fairly extensive.

One common approach is to maintain a fairly detailed record sheet, which contains a number of columns for e.g. hazards, consequences, estimations of probability. The approach is similar to many others in this respect. What makes it "preliminary" is that identification and going through the system may be random, and important aspects might get lost. Like "What-if " described below, its utility depends very much on the skill of users.

(7) "What-if "

"What-if " analysis is a popular technique employed in processing industry. It is not a specific method with a standardised application, but varies according to the user. The basic idea is to pose questions such as:

- **What** happens **if** Pump A fails?
- **What** happens **if** there is an interruption to the electrical power supply?
- **What** happens **if** the operator opens Valve B instead of Valve A?

If the right questions are posed to a skilled team, good results can be obtained. But success using this method is very much dependent on the extent to which the approach is systematic and on the skills of the users. This method can also be fairly extensive.

(8) Coarse Energy Analysis

An Energy Analysis (see Chapter 5) can easily be simplified to provide a simple hazard survey. Simplification involves dividing the system into just a few sections and only considering energies that can lead to fatal or serious injuries.

(9) Coarse Deviation Analysis

Deviation Analysis (Chapter 7) can also be simplified. This means that the division into functions (the structuring) is done rather crudely. Only deviations with relatively major consequences and important planned changes are considered.

12
Methodological overview

12.1 SUMMARY OF METHODS

There are a large number of methods for safety analysis. Two compilations made at the beginning of the 1980s jointly covered 37 methods (Clemens, 1982; SCRATCH, 1984). Since then, a number of further techniques have been developed.

The previous chapters have described methods for safety analysis in varying degree of detail. In all, this book has referred to around 50 methods, including a few variants of the same approaches, and this chapter provides an overview. This section (12.1) is largely concerned with listing the methods presented. Sections 12.2 and 12.3 compare a number of characteristics of ten selected methods. In the final Section (12.4), there is a discussion of how to choose method/methods among all possible alternatives.

Table 12.1 gives a summary of all the methods in the book as a whole. A selection of ten methods has been made, which are fairly thoroughly described and discussed. There are also several methods for which only a general description is made. A number of management oriented methods are mentioned (with names and references). Finally, coarse methods are grouped together.

In sum, the number of methods comes to 45. Several of them can be used both for the analysis of systems and for accident investigations (if slight modifications are made). In all, this means that about 50 methods are considered in the overall text.

Table 12.1   Overview of methods by category.

Type of method	Located	Number of methods
Selected methods	Table 12.2	10
General description given	Table 12.3	19
Management methods	Section 11.6	7
Methods for coarse analysis	Section 11.7	9
Total		45

Selected methods

A fairly comprehensive summary has been made of ten methods, which are compared in some detail. This selection was made on the basis that the methods are fairly simple to apply, and that they can be suitable for workplace investigations. It was further intended that the descriptions in this book should be sufficient to make it possible to apply the methods from this text alone. The ten selected methods are listed in Table 12.2.

This does not mean that the other methods are unsuitable for workplace investigations. The choice could be made wider, and include even more techniques. Such a choice would depend on actual situation for analysis, and also on the analyst's preferences and skills.

Table 12.2   The ten selected methods.

Method	Chapter/ Section	Comment
Technically oriented		
1. Energy Analysis	5	Identifies energies that can harm human beings.
2. Hazard and Operability Studies (HAZOP)	8	Identifies deviations from intended design of equipment.
3. Failure Mode And Effects Analysis (FMEA)	11.2	Identifies failures of components (or modules).
4. Fault Tree Analysis	9	Shows logical connections and causes leading to a specified undesired event.
5. Event Tree Analysis	11.2	Analyses alternative consequences of a specific hazardous event.
Human oriented		
6. Action Error Method	11.3	Identifies departures from specified job procedure that can lead to hazards.
Systems oriented		
7. Job Safety Analysis	6	Identifies hazards in job procedures.
8. Deviation Analysis	7	Identifies deviations from the planned and normal production process.
9. Safety Function Analysis (SFA)	10.5	Analyses safety features and their weaknesses in a system.
10. Change Analysis	11.6	Establishes the causes of problems through comparison with problem-free situations.

Methods briefly described

For purposes of orientation, a number of other methods have been described more generally. In these cases, it is probably not possible to conduct an analysis on the basis of this book alone. However, references are given to facilitate the finding of more detailed information. A summary of these methods is given in Table 12.3.

Table 12.3   Other methods briefly described.

Method	Section	Comment
Technically oriented		
Safety Barrier Diagram	10.4	
Cause-Consequence Diagram	11.2	
Reaction Matrix	11.2	
Consequence analysis models	11.2	Several different methods
Human oriented		
Human Reliability Assessment (HRA)	11.3	A group of methods
Technique for Human Error Rate Prediction (THERP)	11.3	
Cognitive Reliability and Error Analysis Method (CREAM)	11.3	
HAZOP—human errors	11.3	Extended to include
Hierarchical Task Analysis	11.4	
Cognitive Task Analysis	11.4	A group of methods
Organisational oriented		
Audits—in general	11.5	Many different approaches
Management Oversight and Risk Tree (MORT)	11.5	
International Safety Rating System (ISRS)	11.5	Commercial product
SHE audit	11.5	
SCHAZOP	11.5	HAZOP on an organisation
Accident analysis		
Accident Evolution and Barrier Method (AEB)	11.6	
Multilinear Events Sequencing	11.6	
STEP	11.6	

12.2 COMPARISON BETWEEN METHODS

What and how to compare

Comparison of methods raises many separate issues. For example, the SCRATCH project (1984) considered 12 different matters. The basis for classification practice varies within the literature (e.g. Suokas, 1985). The evaluations below are primarily concerned with the application of the methods in the field of occupational accidents.

For practical reasons, comparisons have been largely restricted to the ten methods presented in Table 12.2. They have been described in some detail, and are relatively easy to compare in a number of respects. Further methods— selected to increase range of coverage—are listed in Table 12.3, and described and compared fairly briefly.

Different aspects of safety analysis

A number of themes in the ten methods are discussed. They concern:

- Object of study—modelling and structure.
- Identification, what is identified and how.
- Qualitative or quantitative analysis.
- Safety measures.
- Analytical procedure.
- Use of resources.
- Difficulty of the method, to both learn and apply.

Some of these issues are discussed from a more theoretical point of view in Chapter 14. One aspect concerns modelling, both of the system and the way accidents occur. Another issue concerns quality of results, which is related to both choice of method and how an analysis is conducted.

Object of study—modelling and structure

How the object of analysis is treated by the different methods and how its different parts are included in any method are roughly summarised in Table 12.4.

Structuring is divided into four types, according to how the structure relates to the object. In eight of the methods, structuring is an essential element, which needs to be well thought through. Especially for methods 8 and 9, the design of the "model" is an essential part of the analysis. Methods 4 and 5 are not based directly on the object itself, but show how a certain event or problem is related to it. For the remaining methods, the structuring is more self-evident and simple.

One parameter is whether a certain method focuses on technical and physical aspects of the system (T), on the humans (individuals) who work within it (H), or on organisation and management (O).

Table 12.4   Structuring of the object of study using different methods.

Method	Basis for structuring	Type*	Aspect**
Technical			
1. Energy Analysis	Volumes, which jointly cover the entire object.	2	T
2. HAZOP	Physical properties, e.g. pipes and tanks.	2	T
3. FMEA	Technical components or modules.	1	T
4. Fault Tree Analysis	Not based on the object, but in relation to the top event.	0	T
5. Event Tree Analysis	Not based on the object, but in relation to the initial event.	0	T (H)
Human			
6. Action Error Method	Detailed description of phases of work of the operator.	1	H
System			
7. Job Safety Analysis	Elements in an individual's job task.	2	T, H
8. Deviation Analysis	Activities, e.g. the production flow or job procedure.	3	T, H, O
9. Safety Function Analysis	The system of safety functions.	3	T, H, O
10. Change Analysis	Compare two situations according to a set of variables.	0	T, H, O

*Type of structuring
0. No special structuring of the object
1. Determined by the object description
2. Needs development, but partially pre-determined
3. Essential part of the analysis, carefully done

**Aspect of the object
T Technical
H Human action
O Organisation

Two of the methods, Energy Analysis and HAZOP, are purely based on technical system properties. This also generally applies to FMEA, Fault Tree, and Event Tree, but these techniques can have a wider area of application. In the list, there is a method directed at human action—the Action Error Method (AEM). As stated above, there are a number of similar methods, and AEM should be regarded as an example.

Three of the methods involve the adoption of an overall perspective, considering integration of technique, the human being and the organisation. These are Deviation Analysis, Safety Function Analysis, and Change Analysis. By going beyond their standard application, some of the other methods can also be adopted on the basis of such a perspective.

Subsystem or entire production organisation

Some methods are designed to cover a specific type of system (or subsystem), such as methods 2, 3, 6 and 7. They focus on a certain part of an installation—either a specific job procedure or a technical part of the production system.

In methods 1, 8 and 9, the object of study can vary considerably. It can range from a single machine to an entire factory. Such flexibility is an advantage, but sometimes imposes greater demands on structuring. Methods 3 and 10 can, in principle, also be used at a rather high systems level.

Identification of risks and deviations

A further feature of each method is how hazards are identified. Most of the methods involve consideration of deviations in one way or another. This aspect is summarised in Table 12.5.

One important issue is how deviations are discovered. For several of the methods checklists are available as aids for identification. These are of two types:

1.   Checklists of types of deviations.
2.   Checklists of what can deviate.

Checklists of types of deviations are used in the first four methods on the list. In HAZOP, for example, attention is restricted to physical parameters, thereby allowing an efficient list of types of deviations to be constructed.

Checklists of what can deviate are applied in the two methods Deviation Analysis and Change Analysis.

Four of the methods utilise a "binary" classification; i.e. a deviation is defined and is assumed to be capable of either occurring or not occurring. Binary classification is most evident in the cases of methods 3 and 4, but also applies to methods 3 and 6. Using the other methods, deviations are not defined and classified in such a rigorously binary manner.

Table 12.5 Applications of deviations and a summary of aims.

Method	Deviation from:	Type*	Aim**
Technical			
1. Energy Analysis	Control of energy, or function of barrier.	–	H
2. HAZOP	Intended system function.	1	H
3. FMEA	Technical function of component or module.	1	H
4. Fault Tree Analysis	Normal system function.	–	M
5. Event Tree Analysis	Functions in restoring the system to a safe state.	–	M
Human			
6. Action Error Method	Specified phase in job procedure.	1	H
System			
7. Job Safety Analysis	Safe way of working.	–	H
8. Deviation Analysis	Planned and normal production process.	1 & 2	H
9. Safety Function Analysis	Efficient normally working safety functions.	–	S
10. Change Analysis	The accident-free system.	2	C

*Type of checklist
1. For identifying types of deviations
2. For discovering what might deviate

**Aim
H   Hazard identification
M   Accident modelling
C   Structured comparison
S   Safety characteristics

Aim and analytical procedure

Analytical procedures can be classified in a variety of ways (e.g. SCRATCH, 1984; Suokas, 1985). One possibility is to base classification on the aims of the analyses:

H   *Hazard identification.* The aim is to discover which hazards (undesired events) might occur. This means that efforts are made to identify conditions that might lead to certain kinds of hazards.

M *Accident modelling and accident event description.* How an accident might conceivably occur and what its consequences might be are described in formal terms.

C *Structured comparison.* A comparison is made with some type of "normal system".

S *Assessment of safety characteristics.* The aim is to describe and assess safety features of a system.

The results of such a classification are shown in Table 12.5. The aim of the first six methods is hazard identification, through the adoption of a specific analytical procedure.

Fault Tree Analysis and Event Tree Analysis are examples of methods that involve accident modelling. Such a procedure is more iterative by nature. This means that steps in the analysis cannot be taken and completed one at a time. Instead, it is necessary gradually to work forward to reach a final result.

The third category covers examples of structured comparison. In the shorter list, Change Analysis falls into this category, and a non-accident situation is compared with one where an accident occurs. A number of other methods belong to this category, such as "Audits—in general", MORT, ISRS, and the SHE audit.

Quantitative and qualitative analyses

One common division is into quantitative and qualitative methods. Quantitative might mean that probabilities for an accident are estimated. Both Fault Tree Analysis and Event Tree Analysis belong to the former category, but they can be used without calculating probabilities. The rest of the methods on the shorter list are qualitative (usually simply defined as "non-quantitative").

To the quantitative category also belong several variants of consequence analysis. Probability and consequence estimates can be combined into a single measure, e.g. the probability of dying from a poisonous gas release at a distance of 1 kilometre from an installation.

Systematic support for safety measures

In principle, all the methods provide a basis for the generation of safety measures. In Deviation Analysis and Energy Analysis, the descriptions provide a set of rules, which provide assistance in generating ideas for safety proposals.

12.3 OTHER METHODOLOGICAL ASPECTS

There are a number of different aspects to be considered in selection of methods. Five general ones are discussed below.

1. Level of detail.
2. Time needed.
3. Need for information on the object.
4. Skills available and method difficulty.

Level of detail

If only an overview of hazards is required and the level of ambition is low, some form of "coarse" analysis can be used. Energy Analysis and Job Safety Analysis are also rather quick and simple methods.

Examples at an "intermediate level" are Change Analysis, Deviation Analysis and Safety Function Analysis, all of which can be employed with varying degrees of detail and accuracy.

For detailed analysis of a specific system, Fault Tree Analysis, FMEA or HAZOP can be selected. The longer list (Table 12.2) shows methods that can be used at either a detailed or sometimes intermediate level.

Time needed

The time required to conduct an analysis depends on a large number of factors, such as:

- The object—size, complexity, type of system.
- The analyst—skill, familiarity with the method, etc.
- Availability of information.
- Accuracy requirements.

It is sometimes hard to distinguish the impacts of these factors from the intrinsic characteristics of a particular method. Nevertheless, Table 12.6 provides a rough picture of the time required to conduct an analysis. It is assumed that application is typical, and that the object is of limited size and complexity. The methods have different areas of application, and the systems are not fully comparable.

The quickest methods are Energy Analysis and Job Safety Analysis. Also, a Deviation Analysis can usually be conducted in a relatively short time, provided the level of ambition is not too high. Some examples of specific analyses, showing time taken, are presented in Chapter 15.

Need for information on the object

One key need for an analysis is information on the object. For some methods, a detailed object description is needed (marked "3" in Table 12.6).

Four of the methods can give meaningful results, even if the available information is rather general (e.g. in a design situation). These are methods 1, 7, 8 and 9. Most of the techniques also work at an intermediate level of detail.

It should be noted that systems vary with regard to the extent that activities are regulated and the freedom offered to operators. Methods requiring detailed information might work less well when they are applied to relatively unstructured activities.

Table 12.6   Time for analysis and information requirements.

Method	Time for analysis*	Comments	Information needed**
Technical			
1. Energy Analysis	4h (1h)	Quick scan possible.	1, 2
2. HAZOP	1–2w	15 min per pipeline. 2 h per main unit.	3
3. FMEA	1d–2w	E.g. 30 min per component.	3
4. Fault Tree Analysis	1d–4w	Depends much on the system.	2, 3
5. Event Tree Analysis	4d–2w	As above.	2, 3
Human			
6. Action Error Method	2d–1w	Can be large (several similar methods).	3
System			
7. Job Safety Analysis	2h–2d	5–10 min per work phase.	1, 2
8. Deviation Analysis	4h–2d	Varies according to degree of detail.	1, 3
9. Safety Function Analysis	1d–1w	As above, can be done as a quick analysis.	1, 3
10. Change Analysis	2d–1w	Depends much on whether information is available.	2, 3

Notes:

*Time needed: rough indication, depends on many factors.

Time units: min = minute, h = hour, d = day, w = week.

**Information needed on the object: 1 = General description or oral report.

2 = Comprehensive information, but not all details, 3 = Detailed information.

Skills available and difficulty of the method

The methods vary in terms of the skills they require, and the time they take to learn. Again, it is difficult to provide general guidelines. The outcome will depend on previous knowledge and other factors. But there are, of course, differences between the methods themselves.

People attending courses on safety analysis have been requested to give their views on how difficult some of the methods are to use. This applies to methods 1, 2, 4, 7 and 8. The students' experience of each method consisted of theoretical instruction of about one hour and exercises lasting around half a day.

Energy Analysis and Job Safety Analysis were considered the easiest, followed by Deviation Analysis and HAZOP. Most participants regarded Fault Tree Analysis as the most difficult.

12.4 ON CHOICE OF METHODS

There are several difficulties involved in making general evaluations. One reason for this is that the results of a safety analysis are determined to a great extent by who has conducted the analysis. Adoption of a particular method will not in itself guarantee good results. Further, there is great variation in areas of application and differences between types of users.

Advantages and disadvantages with some methods

Table 12.7 offers a summary of the advantages and disadvantages of some safety analysis methods. It represents a simplification of a rather complicated situation, and should therefore be treated with caution. It does not contain definitive judgements.

For practical reasons, the summary has had to be short and may be somewhat cryptic. Usually, a number of issues are taken up under each method description.

Multiple factors

Not only the characteristics of the methods themselves are crucial. There are also many factors determining optimum choice between them. Examples are:

- Aim of the analysis.
- Types of systems and hazards.
- Available resources for the analysis.
- Personal skill of the analyst.

Table 12.7 Summary of application areas, pros and cons of safety analysis methods.

Method	Application areas	Positive arguments	Negative arguments
Technical (T)		*Can give "strict" results*	*Human and organisation are missed*
1. Energy Analysis	All types of systems.	Simple method, quick, gives an overview.	Limited analysis of causes.
2. HAZOP	Chemical installations.	Well known, many manuals, straight forward to use.	Time consuming.
3. FMEA	Mechanical and electrotechnical systems. Can be widened.	Well established, international standard.	Time consuming. Many possible failures.
4. Fault Tree Analysis	All types of (technical) systems.	Well established, international standard. Logical summary of causes to an accident. Basis for probabilistic calculations.	(Time consuming.) Difficult. Errors can be concealed. Binary (Yes or No).
5. Event Tree Analysis	All types of (technical) systems.	Well established. Clear picture of sequence of events after a failure. Basis for probabilistic calculations.	Rather difficult. Binary (Yes or No).
Human (H)	*People's actions in systems*	*Human actions are essential*	*Difficult to model and predict*
6. Action Error Method (several similar ones)	Well defined procedures in e.g. process industry.	Straightforward use, rather simple.	Focus on normal process. If instruction is wrong? Many possible failures.
– Hierarchical Task Analysis	Map out the task of an individual, all types of systems.	Goal oriented, well structured description. A basis for further analysis.	Does not support identification of risks (not a "real" safety analysis method).

Table 12.7b (continued)

Method	Application areas	Positive arguments	Negative arguments
System (THO*)	*Also organisation O*	*Organisational activities are decisive*	*Difficult to model and predict*
7. Job Safety Analysis	Defined work procedure for a worker or a team.	Simple to learn and apply, similar to traditional safety thinking.	Too traditional, new hazards not found. Not suitable in automatic systems.
8. Deviation Analysis	All types of systems.	THO, general, works on most systems, simple flexible principle.	Sensitive to structuring, many deviations at different levels
9. Safety Function Analysis	All types of systems.	THO, general, works on most systems. Focus on safety, making it right from the beginning.	Rather difficult, results can be presented in different forms.
10. Change Analysis	All types of systems.	THO, general, rather simple principle.	Based on occurred accidents. Assumes that the original system is safe enough.
– Audits—in general	Check of (safety) management (SM), all types of systems.	Essential to check that SM works. With a suitable checklist, it can be rather easy.	Depends much on the checklist. Can be an empty formality. Difficult to apply on "flat" organisations.
Other methods			
Accident investigation Several methods available.	System where accident occurred, or (better) near-accidents.	Can give deep knowledge. Information from real accidents has greater credibility.	Random selection of a part of the system, not exhaustive. Guilt and liability can give bias.
Coarse methods	*Several methods, in principle all types of systems*	*Quicker and sometimes "easier".*	*Less thorough and less reliable results*
What-if & Preliminary Hazards Analysis	General. The methods are rather similar.	Established names. Some descriptions available. With an expert team leader results can be good.	Unclear definition of methods, a precise specification needed. Easily poor results if applied by non-experts.

*THO = Technical, Human and Organisational aspects

How many methods?

A further question is whether one or more methods should be used. The advantage of employing a second method is that a different approach will find different types of hazards. The argument against is that it takes more time, and the analyst might only confirm what he already knows.

There is usually an overlap of identified risks, as illustrated in Figure 12.1. Section 15.8 offers a comparison, in a specific case, between three different methods and the types of hazards they identify. In this case, the overlap was small, and there were clear advantages with additional methods.

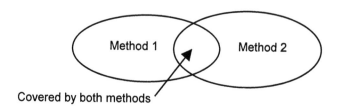

Figure 12.1 Comparison of hazards identified by two different methods.

13
Safety analysis—planning and implementation

13.1 STRATEGY AND PLANNING

Introduction

This chapter points to many situations in which safety analysis can be a useful tool. It is not possible to give a single set of recommendations, which can suit everyone and be universally applied.

The primary aim of this chapter is to give an overview of various planning aspects. With the aid of checklists in table format, it might be possible for a safety analysis user to obtain better results and avoid some of the more common pitfalls. If you make a plan, look through this chapter so you do not forget something important!

The implementation of a safety analysis involves more than the application of one or several methods. Deliberate planning is required for good results to be obtained. When a large installation is to be analysed and many people are affected, planning is of extra importance. If the analysis is limited in scope, planning is simple to carry out. Nevertheless, some of the views expressed in this chapter will still be useful.

Important steps

A short summary list for planning is shown below. It applies in many different situations. Recommendations are to:

- Clarify aim.
- Clarify situation:
  - In design.
  - Existing system.
- Use of safety analysis:
  - On one particular occasion.
  - On a regular basis, e.g. doing an up-date every second year.

- Choice of method:
  - For a first overview.
  - Complementary method. (Are technically, human and/or systems oriented methods needed?)
- Risk assessment:
  - Principle for evaluation.
  - Which risks?
- Organisation of the analysis and social context:
  - External specialist.
  - Internal team.
  - Distribution of competence in the company.
  - Means of communication.

A variety of aspects

This chapter takes up a variety of aspects of the planning, implementation and utilisation of safety analyses. Chapter 14, which is more technically oriented, supplements this description, and takes up sources of error and associated analytical problems. This book contains a number of tables that can be utilised as planning checklists. A list of these is provided in Table 13.1

There is a certain body of literature available on various aspects of the planning of analyses (e.g. CISHC, 1977; SCRATCH, 1984; Suokas and Rouhiainen, 1993).

Table 13.1   Tables related to safety analysis planning.

Table content	Table
Aims of an analysis	13.2
Types of objects analysed	13.3
Situations during system life cycle	13.4
Categories that may be affected by an analysis	13.5
Arguments for safety analysis	13.6
Arguments against safety analysis	13.7
Summary of stages in safety analysis	13.8
Example of a plan for a safety analysis	13.9
Examples of sources of information	13.10
Time requirements for some methods	10.6
Costs and benefits at company level	13.11
Examples of cost-benefit analysis	13.4

There is an increasing number of computer programmes available as aids for conducting safety analyses. Summaries of such programmes easily get out of date, so no attempt to present one is made here. The author's general impression is that there are many programmes available for advanced and large-scale analysis (e.g. for Fault Tree Analysis), but only a few that are applicable to regular occupational accidents. Even when computer support is utilised, it is important to maintain a critical perspective on the analytical procedure and not allow the computer programme to "take over".

Table 13.2   Examples of the aims of safety analysis.

Identification and assessment

Reduce accident risks.

Identification of accident hazards.

Identification of causes of breakdowns of equipment and of interruptions to production.

Identify potential environment problems.

Risk estimation and evaluation (risks and consequences of certain accidents).

Survey of the capacity of the organisation to handle safety matters.

Basis for emergency planning.

Assessment of product safety.

Evaluation of safety in relation to certain criteria, e.g. directives of the authorities.

Finding measures to increase safety

Improvements to machines and layout.

Improvements to job procedures and job instructions.

General improvements to organisational routines.

Provision of assistance to operators at work, e.g. decision support.

Miscellaneous

Meeting the requirements of the authorities.

Integrate safety, health, environment and production into the analysis.

Initiation of a process to promote safety consciousness within the company.

Improvements to and systematisation of safety work.

Provision of basis for decisions to be taken, e.g. for setting priorities or making investments.

Learn how to utilise safety analysis at company level.

13.2 AIMS AND PREPARATIONS

Preparations

Before a decision is made to conduct an analysis, some basic preconditions for success should be investigated. Summaries of a number of different aspects are provided in this section. These are as follows:

1. Aim of the analysis.
2. Type of object to be analysed.
3. Stage in the life cycle of the object.
4. Interested parties.
5. Decision whether an analysis should be conducted, and arguments for and against conducting an analysis.

Aim of the analysis

Table 13.2 provides examples of the various aims that a safety analysis may have. Usually, it is of advantage that several objectives are achieved at the same time. There are a number of advantages to adopting an "integrated approach", so that safety is combined with aspects of production and environmental protection (Section 13.7). One lies in improved finance, as discussed in Section 13.6.

Type of object to be analysed

The nature of the object will determine choice of approach and method. Table 13.3 provides examples of different types of objects, while stages in the life cycle of a system are shown in Table 13.4.

Table 13.3   Examples of types of objects.

Type of system	Comments
Entire installation	Overview required at the start of the analysis.
Type of machine	Standard machine produced in bulk.
Specific machine	Machine tailor-made for the user.
Part of the workplace	Relevant activities and equipment.
Transportation system	Affects technical equipment and routines.
Specific type of work	E.g. carrying out a repair.
Organisational routines	E.g. maintenance routines or planning.

Table 13.4   Examples of situations arising during the system life cycle.

Pre-operation	In operation
Conceptualisation of the system	Normal operation
Pre-planning	Abnormal situation
Planning	Preventive maintenance
Procurement	Repair
Construction	Changes to:
Inspection on delivery	– Technical equipment
System start	– Routines
Disposal	– Staffing
Dismounting the system	– Computer programmes
Clean-up of site	– Products

Stakeholders

People in a variety of different positions will usually be affected by a safety analysis and the changes to which it might give rise. Some may be in possession of important information, or have views on recommendations and decisions. Table 13.5 provides examples of the categories of people that may be affected at various stages of any analysis.

Table 13.5   People who may be affected by a safety analysis.

Within the company	Outside the company
Company executives	Authorities responsible for safety matters
Production staff and job supervisors	Suppliers of equipment
Personnel at the installation	Consultants, e.g. designers
Maintenance staff	Customers
Design staff	Occupational Health Services
Buying staff	Insurance companies
Safety Committee members	The public
Safety representatives	Interest groups
Trade Union representatives	The media

Table 13.6   Arguments for conducting a safety analysis.

	General
1.	Improved safety in the workplace through systematic identification and prevention of accident risks.
2.	Legal requirements that must be satisfied. An overview of risks in the workplace is often a compulsory requirement.
3.	Advantages in comparison with traditional safety work.
4.	In complex systems, the need for a systematic safety approach, so that important aspects are not overlooked.
5.	Safety analysis documentation demonstrating that a systematic approach has been adopted.
6.	Good format for teamwork, because a step-by-step approach is adopted; support for contributions from production personnel on the basis of their experiences.
7.	Improved productivity, if identification of production problems is included in the analysis.
8.	Good business, especially by preventing production disturbances.
9.	A good basis for training and job instructions for maintenance and operations workers.
10.	A good basis for the dissemination of information on hazards and operational safety—to employees, authorities and the general public.
11.	Reduction in insurance premiums.
	At the planning stage
12.	Necessary, if there is a potential for serious accident risks at a new installation, e.g. large consequences.
13.	Safety problems at similar installations.
14.	A suitable opportunity to eliminate hazards and prevent problems.
15.	Cheaper to eliminate hazards and solve problems before they are built into a system.
16.	Basis for improved specifications to suppliers.
17.	Reduction in system-start problems and a shorter start-up period.
18.	Improved understanding of a new installation through systematic analysis.
19.	Good business, supporting opportunities to get things right from the beginning.
	For existing installations
20.	Demand for safety analysis made by an authority.
21.	Serious accident problems and many accidents.
22.	Previous failures in attempts to reduce the number and severity of accidents.
23.	Many incidents (near-accidents) or disturbances to production.

Arguments for and against undertaking a safety analysis

The first step to be taken when planning a safety analysis is to decide whether an analysis should be conducted. Section 1.2 above contains a discussion of the reasons for undertaking a safety analysis. There may be arguments both for and against. Some examples of these are shown in tables 13.6 and 13.7.

There may be reasonable grounds for accepting all the arguments (both for and against). For this reason, they need to be evaluated in the light of a concrete situation.

Table 13.7   Arguments against conducting a safety analysis.

General
1. The level of risk is small, e.g. low "energies", or only office work.
2. Hazards are already known, funds only being required for the necessary safety measures to be taken.
3. There would not be any accidents, if only personnel would follow existing job instructions.
4. Benefits of an analysis are hard to assess.
5. An analysis would be expensive and disruptive to work.
6. An analysis might lead to demands for major and expensive improvements.
At the planning stage
7. Project is far too advanced for a safety analysis to have any effect.
8. Project is at far too early a stage (inadequate knowledge of the form that the installation will finally take).
9. Knowledge of the installation is too poor for an analysis to be undertaken.
10. Supplier has duty to ensure that the machine meets all requirements and is sufficiently safe.
11. Influence on the supplier to make improvements cannot be exerted (lack of willingness on part of the supplier).
For existing installations
12. There is no safety problem, accidents either not occurring or being trivial by nature.
13. The Labour Inspectorate has visited the installation and made no complaints
14. Production facilities may soon be changed.

13.3 ANALYTICAL PROCEDURE

The discussion of planning below is principally based on the model of safety analysis provided in Figure 3.1. Table 13.8 offers a more extensive description of the analytical procedure. It can be used as a checklist before planning an analysis.

Initiation of the analysis

The first step is to make a proposal that an analysis should be undertaken. Someone must take the initiative, and a decision must then be taken. Examples of arguments for and against conducting an analysis are shown in tables 13.6 and 13.7.

Study team and/or management group

It is often preferable to conduct a safety analysis in a team. An extensive analysis can affect a large number of people. A team (or several) can share the work of data collection, analysis, etc. In the case of a major analysis, it may be a good idea to appoint a management group. The group would then participate at important stages of the analysis, such as the formulation of aims, planning and reporting, and also when special problems arise.

A suitable number of team members is between three and six. People with a range of skills should be included, the ideal team composition depending on what is to be studied. All members of the team do not need to be familiar with the analytical procedure.

Several types of problems can arise if just one person conducts the entire analysis. One is that it just results in a stack of paper, which is largely ignored. An important advantage in creating a team or management group lies in the way the analysis can be rooted at company level. Through a stage-by-stage process of clarification and adjustment, results can become broadly accepted within the company. In the course of the analysis, different opinions can be expressed on the way in which risks should be assessed, on the appropriateness of safety measures, on how the system really functions, etc.

The existence of a study team enables the correctness of parts of the results to be checked as the analysis progresses. It is of advantage if differences of opinion are detected early, so that they can be resolved or handled appropriately. Moreover, when people have been involved in conducting an analysis, it is more likely that they will want their conclusions and recommendations to be accepted.

Table 13.8   Summary of stages of procedure in a safety analysis.

Stage	Comments
– Before the analysis	
Initiation	An analysis requires an initiative to be taken, and needs to be rooted at company level.
– Planning	
Appointment of study team	A study team may be needed.
Definition of aims	Objectives are specified.
Setting object limits	How large a part of the object and which activities are to be covered?
Specification of assumptions	What assumptions should be made about the object?
Planning	Schedule and financial resources required. Co-ordination with, e.g., project planning.
Information programme	Those affected by the analysis should receive information.
– Information	
On the system	What information is required? Where can it be found?
On the "problem"	As above. It might concern accidents, disturbances to production, repairs, etc.
– Analysis	
Choice of method	Selection of one or more methods.
Structuring	The object is divided up into sections at an appropriate level of detail.
Identification	Hazards are identified.
Quantification, if needed	Quantitative estimates of consequences and probabilities can be made in some cases.
Assessment	Risks are evaluated. Criteria are needed.
– Safety measures	
Making proposals	Proposals for safety measures are made.
Revising proposals	Ideas are evaluated and worked through.
– Reporting	
Risk summary	Hazards are summarised and commented upon.
Proposals	Summary of safety measures is proposed.
– Decisions	
Decision forum	Who is to draw conclusions and make decisions?
Implementation	Taking decisions is essential.
Follow-up	That measures as specified have been taken is checked.

Definition of aims

The objectives and sub-objectives of any proposed analysis may need to be further defined. This may concern which hazards should be considered, how detailed the study should be, etc. This is discussed in Section 13.2, and an extensive list of various aims that a safety analysis may have is provided in Table 13.2.

Setting object limits

Before starting on any analysis, a number of questions should be addressed: How large a part of the object is to be covered, and what system boundaries are to be specified? Which activities (e.g. maintenance, transport, etc.) are to be included? Will the analysis consider changes to equipment that have been discussed but not yet implemented?

Specification of assumptions

It is necessary to establish what can be taken for granted with respect to how the system functions. The most important assumptions of the analysis should be specified in the final report. Moreover, the correctness of these assumptions should be checked wherever possible. Some examples of assumptions (probably only rarely justified) are provided below:

- Drawings are accurate.
- Job instructions are followed.
- Suppliers of machines will provide adequate solutions to safety problems.
- Machinery will be serviced at prescribed intervals and by skilled personnel. (If quantitative estimates are to be made, this is an important assumption.)
- Personnel have the knowledge and experience to handle unforeseen situations.

Planning

It can be difficult to specify an exact schedule in advance. It is not known how many hazards will be detected and what problems may arise. If such factors are taken into account from the outset, changes of plan need not be regarded as failures. Planning can be seen as an iterative process.

Conducting and planning a safety analysis is simpler in the case of an existing installation. When an analysis is applied to planned plant or equipment, the project schedule for the installation will be the governing factor, and the safety analysis will have to be adapted to this. In addition, access to information will be poorer. Safety analysis in the context of the planning of an installation is discussed in Section 13.3. A simple example of an analysis for an existing installation is provided in Table 13.9.

In some cases, it can be appropriate to prepare a written outline when planning and embarking upon an analysis. This particularly applies if the person in charge of the analysis is not employed by the company in question. Such an outline may include the following items:

- Access to documentation.
- Tasks for different team members.
- Utilisation of results.

Time requirements

The time taken by analyses can vary considerably. It will depend on the nature of the object, the aim of the analysis, the method selected, degree of familiarity with the method, and so on. Estimates of the time required for some methods are provided in Table 12.6.

To take an example, we can look at one of the simpler methods (e.g. Job Safety Analysis, Deviation Analysis, or Energy Analysis) and assume that the team leader has a certain familiarity with it. Information needs to be gathered, but the degree of detail varies. Time required for data collection can be between one hour and several days. Identification of hazards can take between half a day and a full day. These methods also involve a risk assessment stage, which is implemented in one go for all identified hazards. Typically, this might take one or a couple of hours. Generating safety proposals may require a meeting of about four hours, plus some time for further consideration and revision.

An analysis of a medium-sized object using some of these basic methods will require the team leader to devote between two days and a week of his time, possibly more. The study team will meet perhaps three times, each for half a day. But, of course, the time needed may be longer, if the level of ambition is high, or if the situation and object are complicated by nature.

The identification of hazards in itself takes up only a limited amount of time, perhaps 20–30% of that needed for the entire analysis. Some time is taken up in completing record sheets, following up relevant issues and providing information.

Table 13.9   Example of the scheduling of an analysis.

Activity	Date	Respon-sible*	Comments
Preparations	3/1	T	First meeting.
Aims		L	
Setting limits			
Assumptions			
Information to employees	16/1	M1	Workplace meeting, notice board.
Information			Basis for analysis.
Drawings and manuals	27/1	M2	Check for correctness.
Interviews with 2 operators	3–7/2	L	
Reports	10/2	M1	
– Accidents			Check last 3 years.
– Near-accidents			
– Disturbances to production			
Analysis			
Energy Analysis	10/2	L	Designer takes part.
Prepare Deviation Analysis	14/2	L	
Identification of hazards	16/2	T	
Risk assessment	20/2	T	
Safety measures			
Producing ideas	21/2	T	Designer takes part.
Reviewing ideas	24/2	T	Designer takes part.
Summary	3/3	L	List of safety measures.
Time in reserve			
Extra meetings if needed		L	For computer control system.
Decision	10/4	L &M1	Presentation managing director.
Information to employees	12/4	M1	

*Key: L = Leader; M1/M2 = Team members; T = Team participates.

13.4 INFORMATION AND ANALYSIS

An analysis always requires information of different types, both on how the system functions and the problems associated with it. It is perhaps simplest to obtain information by working together with people already familiar with the installation, e.g. through their participation in the study team. Table 13.10 provides examples of various sources of information.

Table 13.10   Examples of sources of information.

On the system
Drawings
Job instructions
Manuals for machines
Maintenance schedules
Computer programmes
Reliability data on components and equipment

On problems
Reports on accidents and incidents
Reports on disturbances to production and breakdowns
Information from other installations
Inspection records
Reports on repairs
Observed damage to equipment, etc.

On both system and problems
Members of the study team
Interviews with designers, etc.
Interviews with personnel at the installation
Direct observations (if installation already exists)
Experiences from other installations
Photographs
Video recordings

Information on the system

Information is needed on the system and how it functions. This may be documented in the form of drawings, job instructions and maintenance schedules. However, written documentation will describe only a limited portion of the reality. First, it will only cover certain system aspects; second, it may not be accurate. Instructions may be out-of-date, may not be followed, or can be simply incorrect.

For this reason, supplementary information is needed. This may be obtained by making observations at the installation. The knowledge of people who know the system can be accessed through interviews, or through their inclusion in the study team. Photographs and video recordings can be utilised, especially for "one-off" operations such as a repair job.

Information on problems

Knowledge of different kinds of problems is of great value, and can be employed in a variety of ways:

- For identifying hazards.
- For judging whether hazard identification has been sufficiently thorough.
- In risk assessment (Have similar hazardous situations arisen before?).
- In discussion over whether safety measures are justified.

Written documentation may concern:

- Accidents.
- Near-accidents.
- Disturbances to production.
- Inspection records of different types.
- Repairs already carried out.

Further information can be obtained through interviews. These can be undertaken in advance or in the course of the analysis. Photographs can be employed to provide documentary evidence that certain problems have actually arisen.

Selection of method

When knowledge of the system and an overall picture of the problems have been obtained, a more definite choice of method can be made (see Chapter 12). It is often a good idea to use more than one method. These can then be used to complement one another by addressing different types of issues.

Structuring

Structuring involves both creating a simplified model of the system and its functions, and dividing the system as a whole into blocks. Several methods require the system to be structured in a specific way. This applies in particular to Deviation Analysis, Energy Analysis, FMEA, HAZOP, and Job Safety

Analysis. Structuring is a critical stage of the analysis and should be carried out carefully (see also Table 12.4 and Section 14.2). Attention should be paid to whether all important elements have been included. This can be done by making checks against the limits of the analysis and assumptions made at the outset.

Identification of hazards

The way in which hazards should be identified is linked to the method employed. It is usually best to follow the structure of the system that has been created and identify hazards for one block at a time.

If hazard identification is carried out in a team, it is important that a creative atmosphere is generated. This means that the team leader should try to avoid criticisms of proposals and assessments at this stage.

In practice, it is sometimes advantageous to relax the planned structure. Interviews and meetings generate ideas and observations that should be followed up before they are forgotten. After these side-tracks have been examined, it is still quite easy to return to the main theme established through the analysis procedure. This means that the team leader must have a grip on the main thread of the analysis, and be able to structure the record sheet afterwards.

Quantitative estimates

If quantitative estimates of probabilities or consequences are encompassed by any analysis, these will usually be made after the identification stage. This provides the basis for selecting the scenarios most relevant to further work. It also gives information that can be used in Fault Tree Analysis, which is the most common technique in this context. Such estimates need expert judgement, and often require access to special computer programmes.

Risk assessment

In qualitative analyses, identified hazards are assessed one at a time. Risk assessment and discussion of safety measures take place when the hazard identification stage has been completed. This can be done even when several methods are employed. The advantages are that assessments are more consistent and a better overall picture of the results is obtained. However, when large systems are analysed, the approach may become unmanageable.

Practical risk assessment is discussed in Chapter 4 (above). Before beginning, it is a good idea to establish which evaluation criteria should be employed. Should disagreement arise, it is not necessary for this to be immediately resolved. Divergent views on evaluation can be noted on the record sheet. These issues will come up later, both when safety measures are discussed and when decisions are made on which measures are to be implemented.

Discussion of safety measures may have an effect on assessment. A greater insight into a problem has been obtained, and it is also known whether the problem can be solved. But, hazards should not be re-evaluated and then regarded as acceptable simply because it is not possible to find a satisfactory control measure. However, the opposite might apply. If a safety measure that can reduce risk is easily available, it is reasonable that it should be implemented.

13.5 SAFETY MEASURES AND DECISIONS

In this section, it is assumed that a list of assessed risks has already been prepared. Alternatively, a fault tree may have been constructed, and control measures are needed to address critical branches. A study team is of benefit, since it facilitates the generation of ideas, and the development and evaluation of safety measures.

Safety proposals

First, an attempt is made to generate ideas for safety measures. The more ideas, the better! Two of the methods (Deviation Analysis and Energy Analysis) contain a systematic procedure for this. If such an open approach can be properly explained, it is easier to obtain a creative atmosphere in the study team. Only at the following stage of the analysis is there a need to exercise greater restraint.

Revision of safety proposals

This stage of the analysis involves a critical examination of proposals. They are reviewed, some are taken away, others are finally revised so that practical and efficient solutions can be obtained. It might not be possible to resolve everything at just one meeting. Some ideas require further treatment.

The following questions should be considered when developing and evaluating ideas for safety measures:

- Does the extent of the measure match the size of the risk?
- What effect will the measure have?
- Does the measure have a short or long-term effect?

- Does the measure have only a local (workplace) effect? Might it only apply to a particular machine or are its effects more general?
- Are there positive or negative side-effects?
- What are the financial considerations? What will the costs be, and how much income will be generated? (See Section 13.6.)

Further investigation

A rather common situation is that specific information about a system or hazard is not available. Sometimes this can be solved quickly. But, on other occasions, it can take a long time—or be costly—to acquire such information. In such cases there are two options, of which the first is to delay completion of the analysis until information becomes available.

In most cases, however, it is better to mention the lack of certainty on the record sheet, and propose an in-depth investigation on that specific point. This saves time, and also helps prevent an excess of obstacles leading to an unfinished analysis.

Reporting

The content of the report generated by the study should largely depend on the aims of the analysis. The report may consist of a risk assessment or a summary of proposed safety measures. An account of the limits of the analysis and its assumptions should be included. The amount of detail required will vary according to situation. For simpler analyses, an oral description and a list of recommended safety measures may be sufficient. In the case of large analyses, reporting requirements can be extensive, e.g. for large chemical installations (CEC, 1996).

Decision-making

A safety analysis should finally result in the taking of decisions. Without a proper decision-making procedure, the efforts made in the analysis can be wasted. The nature of decisions may vary:

- Safety proposals are accepted (and implemented) or rejected.
- Safety proposals are scheduled to be revised, and decided upon later.
- A recommendation for implementation is made, which will be decided upon at a higher level.
- Further analysis of the system is scheduled to take place.
- The analysis, in terms of its quality or the nature of risk assessment, is rejected or accepted.

Implementation

The final stage consists of implementation of the decisions taken. It is of benefit if those who have taken part in the analysis can also participate at this stage. Otherwise, the considerations that lie behind an analysis can be disregarded, with an outcome not as good as it might possibly have been.

Experience of the evaluation of safety analyses suggests that this final link is sometimes the weakest.

13.6 COSTS AND BENEFITS

Introduction

Humanitarian and ethical values play a major role in discussions of the prevention of occupational injuries. In practice, however, great weight is placed on financial considerations. Complete analysis of costs and benefits is complicated. Various types of costs are involved, different parties bear the burden, etc. (see Section 1.1). How costs are distributed in any one country depend on how responsibilities, insurance and social security systems are arranged.

This section considers financial aspects of accidents and safety measures at company level. It describes a simple method for estimating costs and benefits, and provides a brief account of the financial calculations made in a number of case studies. The studies are described in greater detail in Chapter 15.

A cost-benefit analysis is essential, because attention is too often paid to costs alone, which are relatively easy to estimate. The result is that safety measures are often regarded as far too expensive to implement. Benefits and potential financial gains can be large, but they are often intangible and tend to be neglected.

About company costs and benefits

In general, it can be said that the total cost of occupational accidents varies considerably between companies. The costs incurred can be divided into the following categories:

- Insurance premiums.
- Compensation payments, care and rehabilitation of the injured person.
- Production losses, increased production costs, etc.
- Costs of safety measures, safety routines, etc.

A summary of potential costs related to production and accidents might be rather long. Examples are as follows:

- Destruction of equipment and material.
- Interruptions to production, e.g. for accident investigations or repairs.
- Lower productivity and poorer quality in the short term. A substitute worker is likely to be less skilled than the injured person he replaces; work in general may be disrupted.
- Costs of recruiting a replacement and/or overtime payments.
- Indirect disturbances, in that personnel may be emotionally affected by the accident.
- Costs incurred in investigating the accident.
- Specific demands for changes as a result of the accident.

Thus, an accident can be regarded as having a cost with many components. The major financial benefit of safety analysis lies in reducing this cost.

Model for the calculation of costs and benefits

In order to get an economic overview, a simple method for estimating costs and benefits can be employed (Harms-Ringdahl, 1987b, 1990). The calculations are based on costs and benefits at the time when changes are introduced—the "investment". Financial impact on production, etc. during the lifetime of the investment is also estimated—"operations".

Table 13.12 gives a summary of the elements in the analysis, and also includes examples of different types of costs and benefits. The activity "safety analysis" is presented separately since it is intrinsically relevant to safety analysis evaluation.

The principle underlying such economic appraisals is that an original (old) system is compared with a changed system (the new one resulting from applying the results of the safety analysis). The "original" system may be an existing installation or a design proposal. The financial outcome is obtained by assuming a discount (interest) rate and an investment life.

The model can be utilised in different ways. One application is to enlarge the base for final decision-making, i.e. on whether proposed changes should be implemented or not. The items on the list are then studied, and attempts are made to estimate their monetary values.

Alternatively, a cost-benefit analysis might be conducted after measures have been implemented—to establish, for example, whether it is profitable to devote resources to safety analysis. In some cases it is relatively easy to couch estimates in monetary terms, in others more difficult.

Table 13.11 Costs and benefits of safety analysis at company level.

Activity	Comments
INVESTMENTS	During analysis and improvements.
Safety analysis	
– **Costs**	
Time	Mainly working time.
Other costs	For documentation, travelling, etc.
+ **Benefits**	
Knowledge	A better overall picture of hazards and a basis for more rational decision-making.
Training	Of participants; knowledge can be applied in other contexts.
Imputed benefits	Even without an analysis, resources would still have had to be devoted to the problem.
Investments in the installation	
– **Costs**	
Equipment	Implementing technical proposals.
Planning, etc.	Costs of production of manuals and training, etc.
Production losses	Halts to production while changes are implemented.
+ **Benefits**	
Cheaper technical solutions	Cheaper solutions than those specified in original plans.
Shorter start-up period	Shorter run-in times, e.g. through prevention of identified problems.
More rational planning	Fewer changes to the installation after it is ready, e.g. to meet safety requirements.
OPERATIONS	When changes are in place.
– **Costs**	
Reduced production	Slower production runs as a result of the introduction of safety measures.
Other costs	E.g. increased maintenance costs for safety devices.
+ **Benefits**	
Fewer accidents	The monetary value of fewer accidents, largely dependent on the calculation model employed.
Improved work conditions	Improved job performance, reduced risk of human error, less absenteeism.
Improved production efficiency	Fewer disturbances, which also can be handled more effectively, etc.
Reduced risk of disturbances	Reduction in the likelihood of breakdowns and production standstills.

Investment—the safety analysis

The costs of the analysis itself consist primarily of the working time it takes, plus certain extra costs for obtaining information. The benefits lie principally in improved knowledge and a better basis for decision-making. Usually, it is difficult to attribute monetary values to these benefits.

A further benefit of a safety analysis is that it can raise the skills of members of the team that carries out the study. This can be compared with what it would cost to obtain the same improvement by other means. Another conceivable effect is that it improves the expertise of designers and other personnel, whose ability to anticipate and prevent various kinds of problems is increased. Moreover, there may be savings, in that the analysis acts as a substitute for investigations that would otherwise have been necessary.

Investments in the installation

Making an investment involves costs. These may be for the purchase of equipment, or for time spent on design, etc. Administrative costs may be incurred for the production of manuals, job instructions, etc. Further, making changes may give rise to production losses; plant may have to be at a standstill for a time, or the implementation of an investment project might be delayed.

But a safety analysis can also provide benefits in the form of savings— especially when it is conducted at the planning and design stage. When hazards are discovered on the drawing board, changes are easier to make. Planning will be more rational if risks are treated at the same time as technical solutions.

Technical facilities and production methods cheaper and simpler than those originally envisaged might be discovered. Run-in times can be reduced if start-up problems are anticipated.

Operations

There can be costs to making system changes for safety reasons. It may be that the rate of production is reduced, or that a speed limit is imposed on transportation vehicles. Increased maintenance requirements may also entail greater costs.

Table 13.12 also contains four examples of benefits that might result from system changes induced by a safety analysis. The first is related to the basic aim of any such analysis—to have fewer accidents. A reduction in the number of accidents has a financial value. The start of this section provides some examples of the costs to a company that should be taken into account in the calculation.

Work conditions may be improved and task-related problems solved, which can lead to improved job performance and higher work quality. This can also give rise to a reduction in absenteeism, both through reducing the risks of specific occupational diseases and via a general improvement in well-being.

The improved production efficiency resulting from a system which has been well thought through means higher productivity.

Risks for production disturbances can be identified. If their occurrence can be reduced, or ways of handling them improved, financial gains will accrue.

A reduced probability of breakdown can be attributed a monetary value. For example, it may be possible to assume that the frequency of failure is reduced from one in five years to one in ten. If the costs of breakdown can be estimated, it is simple to calculate the resulting monetary amount. Such calculations are always uncertain, but can still be used to provide significant estimates.

Externalities and spin-off effects

Fewer accidents also have a so-called "external" value, i.e. a value that cannot easily be measured in monetary terms. Take the example where a company makes an effort to live up to its safety policy. Demonstrating to employees that the company will take action to deal with occupational hazards improves labour relations and promotes greater job commitment. For there to be such effects, it is necessary that something is really done and that the company's striving to achieve results is made known to its employees.

All this may have a clear economic value, in particular when a customer imposes ethical norms on its supplier. This is now rather common in the environment arena, but also increasingly applies to work conditions.

Another example concerns safety of products sold. A bad reputation for safety can deter the public from buying a particular product.

Appraising the costs and benefits of a safety analysis

Performing a cost-benefit appraisal involves many assumptions and some guesswork. In some cases, data are available on standstill periods, quality problems, accidents, etc. These can be employed to make cost estimates.

Different analysts may come to very different conclusions. But even if there is a great deal of uncertainty, such appraisals are of value when used judiciously. One way of handling the uncertainty is to present results in the form of two monetary values—representing the upper and lower limits of the financial effects of a safety analysis.

Investment appraisal

Investment appraisal involves comparing the immediate expenses incurred in making an investment with prospective earnings that will accrue over a number of years. The profit (P), *calculated at time of investment*, can be expressed as:

$$P = R + C \cdot Y \qquad (13.1)$$

The parameters involved in any such calculation are:

R   Net investment (savings – costs)
Y   Average annual yield (earnings)
C   Factor for calculation of current capital value
n   Life of investment, number of years
r   Discount rate

The factor C is used to calculate the yield of the investment over its entire life. Inflation effects are not taken into account.

$$C = [(1 + r)^n - 1] / [r (1 + r)^n] \qquad (13.2)$$

Table 13.12 *Factors for conversion of annual earnings to current capital values.*

Discount rate	3 years	10 years
5%	2.7	7.7
10%	2.5	6.1
15%	2.3	5.0
20%	2.1	4.2
25%	2.0	3.6

Note that an appraisal involves making a comparison. In this case, it is between the old system as it was before the safety analysis was conducted and the new system that results from the analysis. The net investment (R) is usually negative, but it can be positive. For example, the safety analysis provides a basis for savings during the design of an installation.

Some examples

Cost-benefit appraisals were conducted for five of the examples of safety analysis presented in Chapter 15 below. The principle has been to ask people at the company of their estimates of costs and benefits. Calculations have then been made on the basis of these figures.

There are uncertainties in the estimates, and they depend on a number of assumptions. For this reason, two possible appraisals were made in each case.

Alternative I represents a cautious estimate, whereas Alternative II is rather more speculative with regard to both earnings and expenses. As the examples show, results can vary considerably.

The results of the five appraisals are summarised in Table 13.13. The discount rate was set at 10%, and the investment period was assumed to be ten years (except in Case E where it was three years).

Table 13.13 Five examples of estimated costs and benefits (values in 1000 US dollars).

Example	Alter-native	Net cost (a)	Earnings (b)	Profit	Cost of analysis
A. Planning at a paper mill (Section 15.2)	I	– 90	+ 270	+ 180	– 10
	II	+ 300	+ 340	+ 640	– 10
B. Purchase of packaging equipment (15.3)	I	0	0	0	0
	II	+ 65	+ 60	+125	+ 5
C. Automatic materials handling system (15.4)	I	– 10	+ 80	+ 70	– 5
	II	– 10	+ 200	+ 190	– 5
D. Workplace for ceramic materials (15.5)	I	– 30	+ 330	+ 300	– 3
	II	– 30	+ 1 200	+ 1 170	–3
E. Accident investigation routine (15.6)	I	– 7	+ 15	+ 8	– 3
	II	– 13	+ 75	+ 62	– 3

Comments:
(a) Net investment cost (including safety analysis).
(b) Earnings (based on flow of discounted values).

Figures are given in US dollars, and are indexed at Year 2000 prices. The values should be regarded as indicative, since index-based calculations depend on assumptions concerning production costs, pay levels, and so on. These change over the years and differ between countries. All figures have been rounded so as not to give any false impression of accuracy.

Some plus signs appear in the net-cost column (for investment and conducting the safety analysis). Note here that the calculation reflects the difference between *conducting* and *not conducting* a safety analysis.

For example in Case A, the net-investment cost is a saving of 300 000 US dollars. This is explained by the fact that the cost of constructing the new installation was reduced by decisions resulting from the safety analysis. The differences between estimates I and II are generally rather large, and reflect the different assumptions employed in the calculations. This issue is taken up in greater detail in Chapter 15.

Profitability

Conducting a safety analysis had a favourable effect on company finances in all five cases. This finding does *not* reflect a deliberate decision on the part of the author only to present analyses with a positive economic result. The author has experience of a number of other appraisals, albeit less well documented. Most appear to be economically favourable for the company concerned.

In all the cases exemplified, the primary purpose of the analysis was to analyse and reduce accident risks. However, it is only in the last example (E) that major economic gains accrued from reducing the number of accidents. In this case, the entire economic benefit of the analysis lay in accident reduction.

In the four other cases, the major economic benefits are related to production. This provides a strong argument for adopting an overall approach to safety analysis, i.e. an integrated approach that encompasses both the work environment and production. It means that good proposals can be backed by powerful financial arguments.

13.7 INTEGRATED APPROACHES

Company management aim to attain a variety of different goals. As well as maintaining profitability through the quantity and quality of production, they must satisfy the requirements of the authorities and meet the wishes of their employees. Thus, they must be able to cope with a broad spectrum of problems.

The advantages of an integrated approach to risk issues have been discussed earlier in this book (e.g. in Section 4.4). The concept of an "integrated approach" is understood here as a consistent way of including safety, health, environment and production aspects in management systems. This combination is often referred to by the acronym SHE (Safety, Health and Environment), or by SHEP (if Production is included). This does not necessarily mean that management systems are entirely combined, but there is at least good cooperation.

Integration might concern identification of hazards and problems, risk assessment, reduction of causes and other improvements, and common features in a management system. There is a large spectrum of potential problems, and several may have common or similar causes. Examples of different types of consequences include:

- Occupational accidents and diseases, high absenteeism.
- Fires and explosions.
- Production shortfalls and quality problems.
- Damage to equipment.
- Environmental damage.

Examples of causes include:

- Technical failures especially when combined with inadequate maintenance routines.
- Various types of human error.
- Lack of knowledge and motivation among personnel.
- Inadequate correction of detected problems.
- Poor management solutions and routines.
- Inadequate specification of requirements at the planning and design stage.

Management aspects

Both formal and less formal management systems for SHEP have several similarities. For production quality there are the ISO 9000 standards (ISO, 1987), and for the environment there are standards such as ISO 14000 (ISO, 1996) and EMAS (CEC, 1993). In the case of the work environment there are similar standards and regulations (e.g. Health and Safety Executive, 1991; BSI, 1996). They employ the same key words and phrases, such as responsibility, policy, control system, documentation, and follow-up.

An integrated approach to safety management may also have clear financial advantages. Most of the examples given in Chapter 15 demonstrate that the economic gains from safety analysis do not generally lie in improved safety but in improvements to production. This shows the importance of adopting an overall perspective, even when the only objective of the safety analyst is to reduce the risk of occupational accidents.

Choice of methods

When conducting a safety analysis, an integrated approach does not require the adoption of any peculiar type of method. Several of the methods described above can be used with reasonably good results, but the scope of the identification stage needs to be widened and also include disturbances to production and environment damage. The author has good experiences of applying, say, Deviation Analysis in this way.

At the evaluation stage, it is possible to apply an integrated perspective as described in Section 4.4. Production and quality problems can sometimes be evaluated in financial terms.

One advantage of safety analysis is that it places extra emphasis on how the human being will act within a system, and also takes account of a variety of organisational aspects. This approach can be compared with more common ways of examining production systems—where systems are often optimised in a technical sense, but other features tend to be neglected.

14
Theoretical aspects

14.1 INTRODUCTION

The aim of this chapter is to take up certain aspects of safety analysis in depth. The first theme concerns models and theories related to safety analysis.

The second major topic is the quality of analyses and various factors that might lead to deterioration in results. It is divided into three parts:

- Quality of safety analysis.
- Sources of error in safety analysis.
- About quantitative or qualitative assessment.

14.2 ON MODELS AND THEORIES

General

Models and theories are essential features behind safety analysis in several respects. Aspects include:

- The model of reality on which an analysis is based.
- The model of how accidents occur, which determines the search for hazards and triggers.
- A model of how risks can be controlled.

Many alternative ways of describing systems and their safety characteristics exist. There is discussion over how complex or simplified a model should be.

About models

Wahlström (1994) has published an interesting essay on models in risk analysis, from which several notions have been taken up here. The focus is largely on quantitative analysis, but the discussion is also valid for other types of analysis.

A dictionary definition of a model is that it is a "simplified representation or description of a system or complex entity, especially one designed to facilitate calculations and predictions" (Collins English Dictionary, 1986).

On this perspective, there are three categories of models (Wahlström, 1994):

- *Verbal models* use spoken language and its inherent logic engine. Such a model is often based on *if–then* statements.
- *Symbolic models* consist of a set of symbols and a set of rules how these symbols can be combined.
- *Numeric models* are used to calculate quantitative values.

Deterministic models always give the same output, when a specific input is applied. An essential concept in the modelling process is often causality. The Input U applied to the Real System S gives Output Y.

Wahlström (1994) arrived at a number of conclusions and recommendations concerning models for quantitative risk analysis:

- A model should be refined enough not to be trivial, but simple enough to bring forward only the essential characteristics of the real system. A useful model is a good model.
- Models should be used in a region where they are valid. If a model is used outside its validity region, serious flaws can be introduced.
- Risk analysis and modelling should aim at quantification.
- A model should meet the following requirements: (a) Responses should be repeatable. (b) Predictive power. (c) Based on scientific consensus. (d) Applied theories should be general. (e) Based on a mechanism of cause and effect. (f) Theories should not be contradictory. (g) Minimal number of assumptions.
- The use of deficient models actually poses the most serious threat to the validity of risk assessments.

Model of reality

In the descriptions of method procedures in this book, the term "structuring" has been used. This involves making a model of the object of analysis and showing its different parts. The way modelling is performed is an essential methodological feature. An overview of structuring for ten selected methods is given in Table 12.4.

Five of the methods are largely based on technical system properties: Energy Analysis, Event Tree, Fault Tree, FMEA, and HAZOP. The models become more or less deterministic, and the modelling of a system is fairly straightforward.

Three of the methods in Table 12.4 involve the adoption of an overall perspective, considering integration of techniques, humans and the organisation. These are Deviation Analysis, Safety Function Analysis, and Change Analysis. Especially in the case of Deviation Analysis, a key feature lies in the modelling of functions and activities in the system. The aim of modelling is to assure that all essential functions receive adequate attention. A model can be seen as a kind of a map.

In the overview of methods presented in Table 12.3, the categories Human Oriented and Organisation Oriented contain several methods where modelling is essential, e.g. HRA and THERP. Hierarchical Task Analysis only expresses the aim of constructing a model; further analysis is performed by some other method.

Some of the organisation oriented methods—MORT, ISRS, and SHE-audit—treat modelling in a different way. The methods contain a more or less fixed model as a starting point; the real world is then compared with that model and evaluated.

Models of accidents

There are a number of explanations for accidents that provide a model of how hazards arise and what their causes might be. This was discussed above in Section 2.3. The various methods contain a more or less explicitly expressed model of how accidents occur. This model determines how the identification of hazards and problems is made, and also to some extent how structuring is performed.

Table 12.5 gives indications of "model" related to method. An energy accident model lies behind three of the methods, i.e. Energy Analysis, Job Safety Analysis and MORT. Many of the other methods also include an explanation of accidents couched in energy terms.

Technical deviations are seen as causes of accidents in FMEA and HAZOP, while departures from the normal work process are focused upon in the Action Error Method and Job Safety Analysis. A broader perspective is applied in Deviation Analysis, where combinations of deviations in technical, human or organisational circumstances are regarded as lying behind the occurrence of an accident.

Both Fault Tree Analysis and Event Tree Analysis utilise a model of accidents in terms of combinations of binary events. These events are assumed to be capable of either occurring or not occurring. There is a set of logical conditions under which an accident can occur—specific to each system and situation.

Models of how risks can be controlled

Explanations for accidents are closely related to how risks can be controlled. In the text above, the control aspect has come up a number of times (especially in Chapter 10). In the energy model the notion of barriers is clearly pronounced. And in a number of methods, it is more or less clearly understood that avoidance or handling of deviations is the way for risks to be controlled.

In a Fault Tree, the AND gate symbolises a control, which can prevent an accident from occurring. The same applies to Safety Barrier Diagrams.

Rather extensive modelling of safety features at a specific company, but without formalised logical conditions, is applied in Safety Function Analysis.

Models of safety management systems

The way organisations control hazards is of relevance in a number of methods. Five such methods are listed in Table 12.4. Usually, modelling is based on a hierarchically organised system, starting with policy at top level. Alternatively, modelling might be based on more informal structures. An attempt to include this perspective has been tried in Safety Function Analysis. However, a theoretical discussion of modelling of organisations is beyond the ambition of this book.

A framework for analysing safety management systems (SMSs) has been described by Hale *et al.* (1997). The total activity of an SMS can be presented using a consistent descriptive language. The framework can be used to describe and evaluate an SMS, or to assess the completeness of audit tools designed for SMS evaluation. The approach can also be used as a framework for safety practitioners and managers, and as a tool for accident analysis.

Safety management is seen as a set of problem solving activities at different levels of abstraction, and risks are modelled as deviations from normal or desired process. The framework combines the following principles:

- Safety management seen as a set of problem solving activities at different levels of abstraction in all phases of the system life cycle.
- Safety related tasks are modelled using the Structured Analysis and Design Technique (SADT). This shows the inputs, resources and criteria/constraints necessary to produce required outputs.
- Risks are modelled as deviations from normal or desired processes.

14.3 QUALITY OF SAFETY ANALYSES

Scope

There is no intention here to provide a full account of quality issues in relation to safety analyses. The interested reader is referred to the more specialised literature (e.g. Suokas, 1985; Suokas and Kakko, 1989; Rouhiainen, 1992). This section can be used as a basis for examining completed safety analyses. It may also provide a basis for anticipating and preventing problems when such analyses are planned.

Safety analyses have been subjected to criticism with respect to matters of quality, interpretation and use. The bulk of the publications that take up issues of defective analysis focus on applications to major hazards. For example, the criticism may be directed at probabilistic estimates based on unreliable data for the frequency of component failures and human errors, at a lack of completeness in hazard identification (Suokas, 1985), or at problems involved in the estimation of consequences (e.g. Britter, 1991).

Basic quality questions

Considering quality aspects when applying safety analysis is essential. A suitable starting point is the aim of the safety analysis, which usually is to identify the essential factors affecting the safety of an activity (Rouhiainen, 1992). The quality of a safety analysis can be expressed in terms of its fitness for use. This represents the degree to which the safety analysis is appropriate for its specified purpose. Rouhiainen (1992) points to four major questions in relation to quality of a safety analysis:

1. How well has the analysis identified hazards?
2. How accurately are the risks of an activity estimated?
3. How effectively has the analysis introduced remedial measures?
4. How effectively are resources used in comparison with results achieved?

The questions are essential, but usually impossible to answer with any high precision. Since there are a large number of different applications, it is not possible to find any universal measures of quality. The questions above have been addressed at several places in this book, including the final question, which requires some kind of cost-benefit appraisal of the safety analysis itself (Section 13.6).

Evaluation approaches

A basic quality evaluation can be made by examining the analysis procedure, as exemplified in Section 14.4.

It is possible to conceive of a number of other bases on which a safety analysis can be evaluated (e.g. Suokas, 1985; Rouhiainen, 1990). One way is to compare the results of an analysis with its actual outcome, e.g. accidents that occur. An example is presented in Section 14.4.

The evaluation may focus on the accuracy with which hazards are identified, as a function of the method adopted or the skills of the analyst. It may also concern the precision with which probabilities or consequences are estimated.

An evaluation can be performed by comparing different analyses of the same object, carried out by different teams and/or using different methods. One such example is presented in Section 14.5. Yet another approach involves examining the theoretical basis of the analysis, e.g. the way in which estimates of consequences or probabilities are made.

The procedural approach

One basis for obtaining a favourable result consists in good safety analysis procedure—that it is well planned and implemented. This is in agreement with the general standard for quality assurance (ISO, 1987), which is also based on the idea that a suitable procedure is followed and documented. There are also Norwegian and Danish standards for risk analyses (Norsk standard, 1991; Dansk standard, 1993), both of which support quality aspects. They are primarily based on a procedural approach.

14.4 EXAMINING ANALYTICAL PROCEDURE

Scope

The different steps in any safety analysis combine to produce results that are practical and meaningful. A clearly defined analytical procedure is also the basis to obtain a good quality.

This section gives two examples of how a critical examination can be performed. The first follows the different steps in an analysis and problems that might appear. The second starts with a potential accident and traces the procedure backwards to potential deficiencies in an analysis.

Both examples are based on the assumption that the aim of the analysis is to obtain a safe system, i.e. the implementation of safety measures is included.

Examination of the analysis procedure

The point of departure is the analytical procedure (Figure 3.1 and Table 13.8). The various stages involved and the deficiencies that might arise are examined from a procedural perspective (Harms-Ringdahl, 1987a; Rouhiainen, 1990).

Problems and deviations that may arise are noted for each stage, which in principle represents a form of Deviation Analysis. However, the classification of deviations commonly used is not directly applicable.

Examples of different types of problems are listed in Table 14.1. The list is extensive, but still not complete. Far more detailed lists can be prepared for the evaluation of specific analytical situations (e.g. Rouhiainen, 1990).

The summary in Table 14.1 shows examples of problems that can occur in safety analysis—and they are not uncommon. The summary can be seen as providing a general orientation, but it can also be applied practically.

- When a safety analysis is being planned, a check is made of which problems are especially important to avoid. This might cause improvements to the plan.
- The list can also be used to evaluate an analysis after it has been conducted.

However, discussions about problems too early in a potential analysis might discourage people from even embarking on it. It is better to make a reasonable analysis than not to make one at all!

Accidents—despite safety analysis

Even when a system has been analysed and measures taken, accidents may still occur. A rather different approach is to start with a potential accident and then work backwards to find deficiencies in the analysis.

Figure 14.1 shows a tree of different conceivable deficiencies (Harms-Ringdahl, 1987a). In fact, it is similar to a fault tree consisting only of OR gates (which have not been explicitly marked in the tree). A number of the problems referred to in Table 14.1 reappear in the tree.

Often the crucial question in evaluating safety analysis concerns whether the most appropriate method has been chosen. This is indeed essential, but error in choosing method is just one of several possible causes of an inadequate analysis.

Table 14.1   Examples of deviations that may arise in a safety analysis.

Activity	Deviation	Comments
Planning Aim and object limits	Too narrow	Hazards or subsystems not included.
	Too broad	Cannot cope with the analysis, superficial.
	Unclear aim	See above. The work can be difficult to implement.
Planning	Too little time	Tight schedule, incomplete analysis.
	Staff insufficient	Inexperienced analysts, lack of skills, persons not available.
Information programme	Poorly rooted in the workplace	Active or passive resistance.
		Negative attitudes.
Gathering information On the system	Information unavailable	Certain parts of the system cannot be analysed.
	Incomplete information	Risk for serious mistakes in the analysis
	Faulty information	May apply to drawings, instructions, etc.
	Fault in the model	Assumption that the system functions in a certain way.
	Change in the system	Analysis incomplete
On the problems	Information lacking, or incomplete	Under- or over-estimation of risk.
		Low motivation for risk prevention.
		Not an aid to hazard identification.
Analysis Selection of method	Inappropriate	Neglects certain aspects, e.g. organisational activities.
		Too time-consuming, requires excessive resources.
		Too superficial.
Structuring	Object too narrowly defined	Incomplete analysis.
	Parts excluded	Incomplete analysis.
	Division into sections too crude	Incomplete analysis, superficial.
	Too detailed	Too complex, too much time.
Hazard identification	Sections neglected	Random errors.
	Stressful situation	Lack of time, poor team atmosphere, poor creativity, etc.
	Deliberate omissions	Serious risks can be missed.
Risk assessment	Overestimation	Serious risks get lost in a mass of minor risks.
	Underestimation	Important risks disappear.
	Disagreement in team	Takes too long, etc.

Table 14.1 (continued)

Activity	Deviation	Comments
Safety measures		
Proposals	Lacking	Insufficient basis for proposals.
	Inadequate	Ineffectual measures proposed.
Revising proposals	Good proposals excluded	Too expensive, involve too much change, etc.
	Poor recommendations	Take resources from better measures. May give an impression of triviality.
Decisions		
Documentation, presentation	Incomplete, deficient	Poor impression given, proposal rejected.
Proposals	Good proposal rejected	Hazard remains.
	Poor proposals accepted	Take resources from better measures.
	Decision unclear	Uncertainty in implementation.
	Wrong decision forum	Proposals might not be implemented.
	Decision at wrong time	See above. Timing is critical in design stage.
Implementation		
At executive level	Not carried through	Inadequate follow-up.
	– Lack of resources	Skilled personnel lacking.
		Inadequate financial resources.
		Competition from other activities within the company.
	– Unclear responsibility	Due to unclear decision.
	– In breach of decision	A different opinion may be held by the executive responsible.
At practical level	Implementation failure	Inadequate description of measures.
		Inadequate skills.
		System changed, but measures not adopted.
General		
Planning	Plan not followed	
	Delay in implementation	Especially important at the planning and design stage.
	Failure to obtain required resources	
Analysis team	Failure to attend	Information lacking at meeting, difficult to reach decisions.
	Team does not function	Inappropriate composition, lack of motivation.

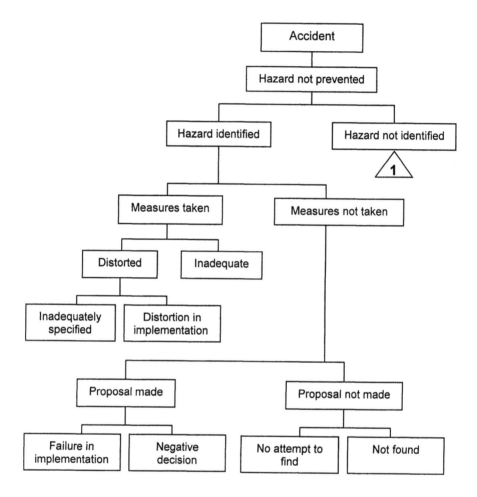

Figure 14.1 Tree for causes of accidents despite safety analysis.

Analysis of a case study

In a case study, a test was made of which types of problem were most common (Harms-Ringdahl, 1987a). The test is based on a case described in Section 15.2, which was followed up at a later date.

Over a two-year period, seven accidents and eight near-accidents were recorded. Explanations for why these occurred are given in Table 14.2—based on a safety analysis perspective.

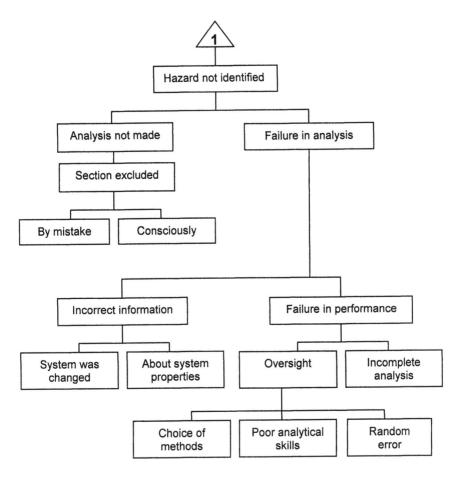

Figure 14.1   (Continued)

The table represents a simplification of the tree shown in Figure 14.1. Some of the accidents can be accounted for in several different ways simultaneously. The selection in the table is based on the principle of starting from the top of the table, i.e. with hazards that were not identified.

There were three major explanations for why accidents continued to occur, each of about the same importance. One problem was that proposals for improvements were not adequately implemented. Another was that some subsystems were not analysed at all. Further, changes were made to subsystems after the safety analysis had been conducted.

It should be noted that none of the cases was categorised as involving an oversight at the identification stage, or as a failure to produce some kind of countermeasure. This does not necessarily mean that these stages of the analysis were wholly successful. There may still be problems, which have not yet manifested themselves in the occurrence of an accident.

A similar test was applied to safety analyses of machines in the paper industry. Identified hazards were related to accidents that had occurred when using a large number of different machines (Suokas, 1985). Of these accidents, 78% had been covered at the identification stage of the analyses.

Table 14.2   An example of the relationship between accidents/near-accidents and a safety analysis.

Explanation	Number	
HAZARD NOT IDENTIFIED		9
Subsystem not analysed	5	
Subsystem changed	4	
Failure to identify	0	
HAZARD IDENTIFIED		6
Proposal distorted	6	
Proposal rejected	0	
Measure not proposed	0	
TOTAL	15	15

Dependence on the analyst

The analyst (team leader) is obviously a key person, who influences the quality of all stages of the analysis. As discussed in Chapter 13, the skills and attitudes of the study team are also important.

Hazard identification is an activity sensitive to the skills of the analyst and the team. The problems that arise in the course of identifying hazards have been examined in several studies. In one case (Suokas, 1985), three analyses of the same object were compared. Only 26% of the identified hazards had been covered by all three analyses.

Dependence on methods

Different methods focus on different hazards and problems. Which aspects are covered by which methods are described in the summary of methods presented in Section 10.1. From this, it is possible to draw conclusions on aspects that can be neglected in an analysis. Selection of method is important for just this reason. As stated above, comparisons between methods have been made, among others by Suokas (1985) and Taylor (1982). It is usually of benefit to employ supplementary methods.

In the case of occupational accidents, one approach is to start with Energy Analysis and then continue with Deviation Analysis or Job Safety Analysis. On some training courses in safety analysis, the results of adopting this approach have been checked against a dozen or so such analyses. Only about one-third of the identified hazards were covered by both the methods employed (represented by the central area in Figure 14.2).

Making the same point in a different way, the employment of a supplementary method will enable roughly 50% more hazards to be identified. (See also Section 15.8, where three methods are compared.)

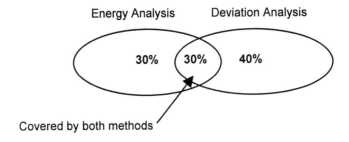

Figure 14.2 Examples of coverage of hazard identification using Energy Analysis and Deviation Analysis.

14.5 PROBLEMS in RISK ASSESSMENT

Introduction

Evaluations of risks are exposed to two diametrically opposed problems. The first is the acceptance of too dangerous a condition, and the second is to demand a change that is unnecessary (in one sense or another).

Different approaches to risk assessments have been discussed in Chapter 4. This section takes up some complementary issues, which are more problem oriented. It also addresses different aspects of quantitative and qualitative assessments.

It is difficult to assess and evaluate accident risks. It cannot be assumed that all analysts will come to the same conclusion. Objective results, those that are independent of the assessor, are impossible to obtain. There is a subjective element to risk assessment, which stems from differences in attitudes and values.

Arguments for and against quantitative assessments

From the 1980s onwards there has been considerable discussion concerning quantitative risk analysis. There are several summaries on this theme (e.g. Taylor, 1994). Some arguments for the use of quantitative risk analysis are:

- The quality of analysis, and of the resulting design, is improved.
- The process of assessment is made much easier
- Assessment of different plants can be made on a uniform basis.

Examples of arguments against are:

- It is costly.
- The applications involve a great deal of uncertainty.
- The methods encourage a rather mechanical approach to plant assessment.
- Although the results can be understood by a layman, they are easy to misinterpret.

Examples of quantification problems

There is often a major problem of uncertainty when quantitative methods are employed. For example, for a number of analyses applied in the chemicals processing industry, a comparison has been made between frequency of failures that have occurred and those that were anticipated (Taylor, 1981). The ratio varied from 0.4 to 1680.

There is an interesting comparison of the different ways in which analyses of the same object can be conducted (Contini *et al.*, 1991; Amendola *et al.*, 1992). In a benchmark study, an ammonia plant was analysed by 11 different teams. Each had the task of conducting a complete risk analysis—from hazard identification to the evaluation of individual risk contours. The results obtained varied considerably. The principal explanations for the differences lay in the following factors:

- Different approaches to risk analysis and its implementation.
- Major variations in reliability data. For certain failures with serious consequences, component reliability data varied by several orders of magnitude.
- Discrepancies in the assessment of human success probability.
- Differences in source term definitions, e.g. with regard to assumptions on how the emission of ammonia takes place.
- Major variations in dispersion calculations.

Probabilistic assessment at "common" workplaces

This book mainly addresses the application of safety analysis to common rather than high-risk, large-scale workplaces. Quantitative estimates of e.g. probabilities can be a part of the risk assessment. There are a number of advantages to this (see the points above).

There are also difficulties with quantitative estimates. Examples include:

1. *Statistical uncertainty.* Even if calculations and data are of high quality, stochastic error in predicting outcome at a particular installation can be large. If an analysis is applied to a mass-produced piece of machinery, an estimate of this sort may be more meaningful.
2. *Estimation uncertainties.* These can be large. At a guess, the uncertainty factor could be about 10 (see comment below).
3. *Information on probabilities.* Data are generally lacking for a variety of types of relevant basic events.
4. *Range of consequences.* The consequences of a certain event can vary considerably. For example, a fall from a height of two metres may lead to very serious injury, but a person might also escape unharmed.
5. *Behavioural adjustment.* People take account of occupational hazards to a greater or lesser extent. There is no unequivocal relationship between the presence of physical hazards and the occurrence of accidents. This is discussed, for example, in risk homeostasis theory (Wilde, 1982).
6. *Systematic errors.* Applying an erroneous model of the cause of accidents, e.g. mistakes in designing a fault tree, can have a major effect on calculations.

Comment on uncertainties

Consider a world median rate of one accident per ten workers a year (Takala, 1998; see also Section 1.1). An uncertainty factor of 10 would give a frequency range between 1 and 100. A rate of 1 would denote a safe workplace, whereas one of 100 would mean that the risk is very high. The limits for acceptable accident risks are generally much more precise.

The problems listed above are related to the accuracy of obtained values. From a workplace perspective, it is clear that probabilistic calculations are time consuming. In most normal workplaces, resources for quantitative estimates are too limited, and the effort involved would probably not be worth the benefit.

15
Examples of safety analysis

15.1 INTRODUCTION

This chapter describes a number of simple examples of safety analyses. The aim of this chapter is to show that analyses can frequently be conducted using relatively simple methods and involving only a small amount of work. The examples have been selected to illustrate choice of method, analytical design, time spent on the analysis, and the results that can be obtained.

The examples concern:

A. Design for the rebuilding of a section of a paper mill (Section 15.2).
B. Purchase of automatic packaging equipment (15.3).
C. Automatic materials handling system (15.4).
D. Workplace for production of ceramic materials (15.5)
E. Investigation of accidents (15.6).
F. An integrated analysis of a chemicals processing plant (15.7).
G. A study of safety functions and comparison with other methods (15.8)
H. A quick analysis of a production line (15.9)

Most of the examples come from the author's own applications of safety analysis, but in cases D and H the work was carried out by safety engineers at the company in question.

Economic appraisals

Economic appraisals were undertaken in five cases. These were performed in accordance with the principles described in Section 13.6. Alternative financial estimates were made in each case. *"Alternative I"* is the more cautious of the two, and relatively low values are assigned to costs and benefits. *"Alternative II"* assigns higher values to some items, based on estimates that are rather more theoretical. This illustrates the uncertainties and difficulties involved in making appraisals of this sort.

In cases A, B, C and E, estimates of costs and benefits were made by people inside the company, in conjunction with the author. In Case D, a safety engineer carried out the appraisal, and based his estimates on information made available by the company in question.

The estimates do not take externalities into account, and are based solely on income accruing to and costs incurred by the company. In Sweden, the compensation costs of the employer are wholly covered by an insurance policy or equivalent scheme. Moreover, the insurance premium is independent of the number of accidents that occur. For this reason the costs of accidents would have been higher if those borne by individuals and society had been included.

A discount rate of 10% and an investment period of ten years were usually assumed. Since the investments were made during several different years, all amounts have been indexed at Year 2000 prices with appropriate currency translations.

15.2 ANALYSIS ON DESIGN OF A PRODUCTION SYSTEM

Background

There had been a high frequency of accidents in one production department of a paper mill over a period of many years. A decision was made to rebuild the section, partly because of high accident risk. In conjunction with rebuilding, safety analyses were conducted of various parts of the section (Harms-Ringdahl, 1982; 1987a).

A description of the analysis and a summary of its economic effects are provided here. Safety analysis was applied on a number of occasions. Short accounts of some of these are given below. They concern the subsystems:

- Layout and materials transportation.
- Paper machine (to be purchased).
- Paper machine (to be remodelled).

Overall analysis procedure

Rebuilding was followed from the pre-planning stage to the completion of the installation, a period of about a year and a half. There was a working group concerned with work environment and safety. The group scrutinised the results of the analyses and produced recommendations for control measures.

Planning was essential so that the analyses could be carried out at the right time, i.e. when proposals of sufficient detail were available, but also in good time before decisions were due to be taken.

On the whole, this was managed successfully. It was possible to apply safety analysis and to apply the results both in design and at purchase of equipment. Problems arose when some parts of the materials transportation system were to be purchased. There was not enough time to conduct the analyses, and subsystems were bought without them having being examined from a safety perspective.

Analysis of layout and transportation system

The conveying of materials to and from machines and the storing of finished products are important parts of operations in a paper mill. On the rebuilding of the mill, a major change was made to the materials transportation system.

Analysis procedure
Layout and transportation were analysed together. Issues were discussed at six special meetings of the work environment group, each of which lasted for about two hours. The discussion was based on a proposal for layout and materials transportation. The proposed arrangement was divided up into suitable sections, each of which was analysed separately. Energy Analysis and Deviation Analysis were employed simultaneously. The analysis was conducted in the following stages:

a. The normal materials flow was checked. Energies and the deviations and problems that might arise were studied for each section.
b. Occasional forms of materials transportation were analysed separately. The analyses were applied, for example, to supplementary materials, and to waste disposal. These work phases tended to involve various types of manual handling.
c. Pedestrian traffic was treated as a special phase of work. Different types of hazards were identified, resulting in a proposal for suitable walkways.
d. Special attention was paid to industrial truck traffic. The trucks were used for both routine transportation and occasional forms of materials conveyance.
e. Maintenance and repairs were affected by the design of the layout. Specific points investigated were accessibility, lifting facilities and aspects of transportation.

The analyses focused directly on immediate improvements. Whenever a problem was identified, the aim was to find a direct solution. Documentation was arranged by directly writing in changes on a drawing. When this document was redrawn, it provided the basis for a new round of analyses.

There was insufficient time for certain parts of the layout to be covered by the analyses. This depended in turn on changes having being introduced, partly as a result of earlier analyses. It meant that certain parts of the transportation system were ordered without there having been an opportunity for the study team to inspect them.

Figure 15.1   Some occasional forms of materials transportation were covered by the analysis. Here, waste paper is being disposed of.

Economic appraisal

The analyses of the transport system led to improvements and changes, some of which were extensive. The most expensive concerned a change to a conveyor belt system. A total cost of around 100 000 US dollars was incurred.

Savings could be made by simplifying the system in several places, and these were valued at 50 000 US dollars. It was originally intended that automatic industrial trucks should form part of the transportation system. This idea was abandoned at the planning stage, resulting in a saving of 500 000 US dollars. This decision was partly based on safety considerations.

The analyses gave rise to extra design work (of about a week), but this was partly compensated for by co-ordination benefits. The length of the run-in period was not affected, either positively or negatively.

The changes did not lead to any increased operational expenses. The changed layout provided more storage space, which was valued at 25 000 US dollars per year. The cost of transportation by truck was reduced by 10 000 US dollars per year due to a more rational materials flow. In total, extra production benefits were estimated at a value of around 35 000 US dollars per year.

The risks of disturbances to production may well have been reduced. Although such gains are difficult to estimate, an example is still provided. The new layout offered better opportunities for work to be carried out when the fixed conveyance system was out of order. A breakdown in the materials handling system might halt production for 24 hours—at a cost of 200 000 US dollars. On the assumption that this could be avoided once every 20th year, the reduced risk can be estimated to have a monetary value of 10 000 US dollars per year.

Purchase of a new machine

A new paper-rolling machine was to be purchased while the section was being rebuilt. It was intended that this should be as safe as possible. The analysis was based on offers made by two suppliers and the study of two similar machines already in operation at other paper mills.

The two machines were examined using Energy Analysis and Job Safety Analysis. Study visits were made to gather information on accidents and near-accidents, disturbances to production, and maintenance problems. These resulted in a specification for safety, maintenance, etc. that was later presented to and accepted by both manufacturers. The specification was later included in the purchasing contract.

Economic appraisal

The eventual supplier originally required an extra 50 000 US dollars to meet new work environmental requirements. About half of this amount concerned safety, the other half noise and vibrations. This extra amount was eliminated during negotiations, and the final price of the machine was not affected. Thus, since the extended specification did not generate any operational benefits or costs, the net cost was set to zero (in Table 15 under heading Alternative I).

On the other hand, without the new requirements, it might have been possible to reduce the purchase price by a further 30 000 US dollars. This is the figure employed in Alternative II.

Remodelling an existing machine

An existing paper-rolling machine was to be moved to a new position during rebuilding. It was originally intended that there should be no changes to the machine, but a number were made for safety reasons.

The analyses were carried out by a study team consisting of three people. Job Safety Analysis was the principal method employed. This covered 36 phases of normal job procedure plus a few extra phases to account for occasional tasks. The analysis took about five minutes per phase.

A special study was made of five job phases where there was a particularly serious risk of a part of the body getting caught between paper reels and cutters. It proved possible to find alternative work procedures, which were expected considerably to reduce the risks.

The risk level at this machine was assessed to be unacceptable for a number of hazards. The analyses resulted in a list of safety measures containing around 50 items. These could be divided up into the following categories:

- Technical arrangements.
- Control systems for the paper-rolling machine.
- Job procedures.
- Inspections (to be carried out more regularly).
- Functional requirements (not specified in detail).

Figure 15.2   A hazardous phase of work is to feed in paper from an already positioned reel.

Economic appraisal
The remodelling of control and hydraulic systems cost around 30 000 US dollars. However, the same kind of conversion would have been needed sooner or later. An extra cost of 10 000 US dollars was incurred by having to transport reels to another paper mill while the conversion took place. This could have been avoided if the work had taken place at a more appropriate time.

Operational costs were not affected. Some employees have maintained that the work takes a bit longer because certain shortcuts are no longer permissible; others think that the finished reels are of better quality.

Changes in the frequency of accidents

The rebuilding of the section led to a reduction in the number of accidents. Information on accidents is shown in Figure 15.3. The rebuilding took place during the first six months of 1982.

A comparison has been made between the four years preceding the rebuilding of the section and the five years that followed (excluding the year when rebuilding took place). The average number of accidents fell by 55%, and the reduction is statistically significant ($p < 0.001$). Absence from work as a result of accidents fell by 40%. In other sections of the paper mill, the number of accidents fell by 20%, while absence rose by 10%.

As can been seen in Figure 15.3, accidents did not disappear completely in the rebuilt department. Possible reasons for why they occurred despite safety analysis are described in Table 14.2.

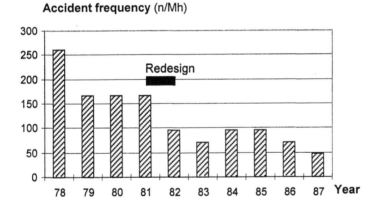

Figure 15.3 Accidents before and after rebuilding (frequencies in accidents per million working hours).

Economic appraisal

The costs and benefits of applying safety analysis were investigated, and a summary of the economic analysis is presented in Table 15.1. (Some comments on the various financial items have been given earlier in accounts of subsystems.) The description is detailed, so as to illustrate how a calculation can be made.

Analysis costs

The analyst devoted about six weeks of his time to work directly related to the project. Meetings and participation in the analysis took up about eight man weeks of the time of personnel at the mill. About the same amount of time would have been expended even if design and planning had been carried out in the usual manner (i.e. without a safety analysis). Some of those involved were of the opinion that the analysis actually saved time, since the design work was conducted with greater efficiency. Thus, it can be said that the safety analysis did not involve extra work for employees.

It was considered that working with safety analysis had an educational effect. Thus, the net cost of the analysis is the difference between the cost of time spent by the analyst and the value of training. The analyses did not give rise to any delays in the project, which was implemented on schedule. For this reason, no such costs have been included in the calculation. The net cost was estimated at 10 000 US dollars.

Table 15.1   Costs and benefits of the application of safety analysis to the rebuilding of part of a paper mill (1000 US dollars).

Activity	Alternative I	Alternative II
INVESTMENTS		
Safety analysis (net)	– 10	– 10
Investment costs		
Layout and transportation system	– 100	– 100
New paper-rolling machine	0	– 30
Remodelling of existing machine	– 30	– 40
Investment benefits		
Layout and transportation system	+ 50	+ 500
Total	– 90	+ 320
OPERATIONS (per year)		
Improved productivity	+ 35	+ 35
Reduced risk for disturbances	0	+ 10
Fewer accidents	+ 10	+ 10
Annual total	+ 45	+ 55

Economic benefits of fewer accidents

The average number of days absent as a result of accidents was 91 per year before the rebuilding of the workshop. This fell to 56 after conversion, meaning 35 fewer days of absence per year. At this paper mill, the cost of replacing an employee was estimated at 200 US dollars per day (the remaining costs being covered by insurance). Including some extra costs, this represented a saving of about 10 000 US dollars per year.

Investment in safety

The application of safety analysis involved extra costs for the transportation system, and the new and remodelled paper-rolling machines. However, ideas for improvements also resulted in a number of savings. According to the more careful Alternative I, extra expenses of 90 000 US dollars were incurred. However, under the more extravagant Alternative II, the investment was estimated to represent an immediate saving of over 320 000 US dollars.

Changes in operations costs

Additional value for production was estimated at around 50 000 US dollars per year. Future earnings can be translated into current capital value using the discounting method described in Section 13.6. Assuming an investment life of ten years and a discount rate of 10%, the conversion factor is 6.1. This gives 270 000 or 340 000 US dollars depending on the alternative chosen.

Total outcome

Applying safety analysis gave a profit of 180 000 US dollars according to estimates in Alternative I. This goes up to 660 000 US dollars in Alternative II. The more careful form of planning and design that safety analysis involves would have been profitable whichever method of appraisal was adopted. However, it would not have been profitable if only savings from a reduced number of accidents had been taken into account.

15.3 PURCHASE OF PACKAGING EQUIPMENT

Background

Equipment for the automatic wrapping of paper reels had been bought by a paper mill. Following purchase, the buying company wanted accident risks to be thoroughly examined. It was decided that a safety analysis should be conducted. The results of the analysis would then be discussed with the supplier and taken into account when the system was manufactured and installed.

The planned installation was complicated. It was controlled by a computer that needed to be co-ordinated with other computerised systems. About 30 mechanical movements were involved, and two materials handling robots were included. To some extent this was a tailor-made system, but a number of similar systems had previously been manufactured.

Figure 15.4. An overall view of the packaging installation.

Analysis procedure

The amount of time available for the analysis was limited and not really sufficient for an installation of such size and complexity. Study visits, each of just over a day, were made to two similar plants. Including written reporting and meetings, the analysis took just over a week.

"Quick" Energy and Deviation Analysis (see Section 11.7) were applied. In the latter case, there was insufficient time fully to follow the analytical procedure. An attempt was made to identify as many deviations as possible through interviews with the people involved.

There were a large number of powerful machine movements. Also, there were several examples of disturbances to production whose correction required work in proximity to the machines. The analysis resulted in a list of hazards and ideas for safety measures. This list was employed in discussions between the paper mill and the supplier.

Economic appraisal

Cost of safety analysis

Just over a week was devoted to the analysis. Had it not been conducted, other consulting services would have been needed. At the paper mill, it was judged that the safety analysis was twice as efficient as a more common form of investigation. In addition, the safety analysis had an educational effect equivalent to one man week of training. Thus, the analysis itself generated a net labour saving of just over two weeks, with an estimated value of 5000 US dollars.

Table 15.2   Costs and benefits of a safety analysis applied to a packaging station at a paper mill (1000 US dollars).

Activity	Alternative I	Alternative II
Investments		
Safety analysis (net)	0	+ 5
Investment costs	0	+ 10
Shorter run-in time	0	+ 50
Total	0	+ 65
Operations (per year)		
Improved quality, etc.	0	+ 10
Fewer accidents	0	0
Annual total	0	+ 10

Investment to improve safety
The changes did not give rise to either extra expenses or time delays for the customer. That demands for improvements arose before the machine itself was manufactured may be expected to have provided further savings, mostly for the supplier. Those accruing to the customer are estimated at 10 000 US dollars.

The run-in period was considered to have been reduced by a month through the prior identification of a number of problems. This was based on a comparison with a similar machine from the same supplier, which had been delayed by several months as a result of run-in problems.

A simple calculation can be made on basis of the value of the installation, assessed at five million US dollars. Assuming that this is depreciated over ten years and that the equipment is available for one extra month, a figure of around 50 000 US dollars is obtained.

Operations benefit
The effect on operations is difficult to assess. Also, it is hard to estimate how the risk of accidents has been affected. In the judgement of the supplier, the careful work conducted by the paper mill at the planning stage provided for better machine accessibility than that for other machines. Employing cautious reasoning (Alternative I), financial costs and benefits might both be assessed at zero.

In the course of the analysis, certain problems related to man-machine interaction were observed. This led to improvements in the workplace, probably resulting in reels of higher quality and better appearance.

As an exercise, a theoretical venture guess can be made. Defective wrapped reels can lead to an order being missed, one perhaps in an amount of one million US dollars as net income. Supposing that the improvement leads to one fewer missing order every 100 years, the annual saving will be 10 000 US dollars per year.

Assuming an investment life of ten years and a discount rate of 10%, the translation factor is 6.1. This would give a current capital value of 60 000 US dollars.

Total result

The financial considerations are summarised in Table 15.2. The cautious estimates of Alternative I result in a zero financial outcome; the safety analysis neither incurs costs nor generates income.

According to the more generous estimates of Alternative II, the safety analysis reduces investment costs by 65 000 US dollars, and has a hypothetical current capital value of 60 000 US dollars. Total profit comes to 125 000 US dollars.

15.4 AUTOMATIC MATERIALS HANDLING SYSTEM

Background

This example concerns an installation for the automated sorting of loading pallets. The pallets are moved into a sorting area and inspected automatically. The ones that are damaged are sorted to one side to be sent on to a repair shop. Those that are wholly intact are stacked, fastened and stored.

The equipment is tailor-made and designed for fully automated operation. It is controlled by a computer, which governs 50 different mechanical movements. Information comes from about a hundred sensors.

The description here is mainly aimed at illustrating cost-benefit appraisal, not so much on how safety analysis is performed.

The safety analysis

The installation was used as a practical example in a training course on safety analysis. On the course, around 30 proposals for safety improvements were made. Before this, a safety engineer and an officer from Sweden's Labour Inspectorate had carried out normal inspections of the equipment. Both persons had considerable experience. At these inspections, six accident-related proposals had been made. This suggests that safety analysis can be a significantly more efficient means of identifying occupational hazards than standard inspection.

Figure 15.5 Part of the automated pallet handling system.

A deeper analysis of the installation was made in conjunction with a later research project (Harms-Ringdahl, 1986). This led to the generation of further safety proposals. In brief, the analyses demonstrated that there were a number of machine movements for which protective devices were lacking. One explanation for this was the assumption made at the design stage that the machines would run fully automatically. Accordingly, it was believed that there would be no reason for anyone to be in proximity to them.

There proved, however, to be a large number of reasons why manual interventions had to be made. There were several possible sources of disturbances to operations, all of which required manual correction. Safety proposals focused on how these should be handled and, to some extent, on how the frequency of such disturbances should be reduced.

Economic appraisal

Aim

The aim of the appraisal below is to estimate what potential benefits or disadvantages there might have been if a safety analysis had been conducted at the planning and design stage. What lessons can be drawn from such a case?

No distinction is made between supplier and customer costs, since the distribution of these would have been a matter for negotiation between them. Although the reasoning employed in this example is rather complicated, it may give the reader an idea on how to proceed in similar situations.

Cost of safety analysis

A couple of weeks would have been needed for the analysis and discussion of safety measures, at an estimated net cost of 5000 US dollars.

Investment to improve safety

Further expenses would have arisen from extra design and construction work, which may be estimated to have cost between 10 000 and 20 000 US dollars.

It may have been possible to implement the project without any significant delay. An alternative assumption is that system start would have been delayed by three months. This would have meant that four employees would have had to continue with the manual handling of pallets for this period—at an estimated cost of 40 000 US dollars.

The run-in time for the installation was fairly long, involving extra work for a number of months. The improved design would have reduced that considerably—providing a conceivable saving of 20 000 US dollars. Changes to the equipment have been made since it was put into operation. Some of these could have been avoided, resulting in a further saving of 5000 US dollars.

Operations

On the operational side, the frequency of disturbances has meant that one person now has to monitor the equipment virtually all the time. If some of these disturbances had been prevented, it might have been possible to make savings equivalent to half the working time of one man. There are opportunities for this time to be utilised for other tasks at the plant. Accordingly, the labour cost of a half-time worker (20 000 US dollars) has been included under Alternative II.

The work station for the person doing the monitoring is poorly designed. Had it been better planned, work conditions would have been better and efficiency greater. It was estimated that 20% of working time was lost as a result of work station design. It would have been possible to utilise this time for other activities, and this has been attributed a value of 8000 US dollars.

The installation has a certain amount of over-capacity, but overtime working has still been needed to compensate for interruptions to production. The cost of this has been estimated at 6000 US dollars per year. It should have been possible to prevent a large number of halts to production.

Table 15.3 Costs and benefits of safety analysis applied to a materials handling system (1000 US dollars).

Activity	Alternative I	Alternative II
Investments		
Safety analysis (net)	– 5	– 5
Investment costs		
Design and construction	– 10	– 20
System delay	0	– 40
Investment benefits		
Shorter run-in time	0	+ 50
Avoidable changes	+ 5	+ 5
Total	– 10	– 10
Operations (per year)		
Improved efficiency	+ 14	+ 34
Fewer accidents	0	0
Annual total	+ 14	+ 34

No accidents leading to an absence from work had occurred at the time of the evaluation, which suggested that increased safety had a low monetary value.

Assuming an investment life of ten years and a discount rate of 10%, the conversion factor is 6.1. This would give a current capital value of 80 000 or 200 000 US dollars depending on the alternative chosen.

Total

Table 15.3 lists the various items included in the estimates. The extra costs involved in applying safety analysis are small in both alternatives. The conclusion based on these results is clear. It would have been of financial benefit to conduct a safety analysis before the project was implemented.

Although there are major uncertainties in the calculations, margins are large, and this conclusion likely to be valid.

15.5 WORKPLACE FOR PRODUCTION OF CERAMIC MATERIALS

Background

For a long time, there had been problems at a workplace where chemicals were mixed for the production of ceramic materials. A lot of manual work was involved, some of which was very heavy. Moreover, a large amount of quartz dust was generated by the work. A few years earlier, the safety engineer at the company had conducted an investigation and proposed improvements. The company rejected the proposal on grounds that it was too expensive.

But the problems remained, and the safety engineer was requested to come up with proposals to reduce high levels of sickness absenteeism and personnel turnover at this workplace.

The description here is mainly aimed to illustrate cost-benefit appraisals, and less to discuss how the safety analysis was performed. The various items in the appraisals are indicated in Table 15.4. However, a detailed report of these is published (Harms-Ringdahl, 1991).

The safety analysis

On this occasion, the safety engineer carried out a safety analysis supplemented by an economic appraisal. Energy Analysis and Deviation Analysis were employed. The entire investigation took about a week, of which around one day was devoted directly to the safety analysis.

The analysis uncovered many reasons why an accident might occur. There were a number of high-energy sources, and many possible deviations in procedures. The analysis also highlighted ergonomic difficulties, largely associated with overexertion due to the lifting of heavy sacks and the accentuated problems of exposure to quartz dust. Many of the ergonomic and hygienic problems were already known.

The analysis resulted in a list of suggested improvements to the workplace. The overall proposal was even more extensive than the one earlier rejected by the company.

Economic appraisal

Aim

Since the earlier proposal had been regarded as too expensive, it was natural also to estimate its benefits. The suggested improvements were supplemented by a cost-benefit analysis, and the estimates are summarised in Table 15.4.

Investment
The cost of investment applied to rebuilding of the workplace, and also safety analysis and the investigation.

Improved work conditions
In the table, the first three items under "Operations" are concerned with improved work conditions. Assessments were made of potential reduction in costs due to overexertion injuries, absenteeism, and personnel turnover.

Improved production
The safety analysis revealed that there were several possible sources of human error, which would result in a faulty mixture. When this was followed up, it emerged that such errors had been made several times a year but that production management was generally not aware of these.

One consequence was that when a fault was detected, the mixture was simply disposed of. Usually, it was simply poured down a drain, also creating an environment problem.

If the error was not detected, a faulty mixture was introduced into the production process. This might lead to major disturbances to production at a later stage. The costs were difficult to estimate, but they were allocated a value of 150 000 US dollars per year in the calculation for Alternative II.

The benefits for production are that the likelihood of such errors is reduced, and also better cared for when they arise.

Table 15.4   Costs and benefits of improvements to a workplace for the mixing of chemicals (1000 US dollars).

Activity	Alternative I	Alternative II
Investments		
Safety analysis and investigation	– 3	– 3
Technical changes	– 30	– 30
Total	– 33	– 33
Operations (per year)		
Reduction of overexertion injuries	+ 6	+ 6
Reduction of absence	+ 8	+ 8
Reduction of personnel turnover	+ 10	+ 10
Reduction of wastage of materials	+ 30	+ 30
Reduction of production disturbances	0	+ 150
Annual total	+ 54	+ 204

Operations

The estimates are summarised in Table 15.4. According to Alternative I, possible reduction of losses to production were estimated at 54 000 US dollars per year; in the case of Alternative II, they were assessed at 204 000 US dollars.

Assuming an investment life of ten years and a discount rate of 10%, the conversion factor is 6.1. This would give a current capital value of 330 000 or 1.2 million US dollars depending on the alternative chosen.

Results

There would be clear economic advantages in implementing the changes suggested by the safety analysis. The investment would be paid back in less than a year.

When company management was made aware of the benefits of improvements (on the basis of Alternative I), an immediate decision was made to redesign the mixing facilities.

15.6 ACCIDENT INVESTIGATIONS

Background

As part of a research programme, experimental safety routines were introduced at two paper mills for a period of around six months (Harms-Ringdahl, 1983; 1990). The aim was to investigate what could be achieved by applying a systematic methodology for accident investigation as a part of safety activities in the workplace.

The usual procedure was that a simple accident investigation was carried out. This was performed by the job supervisor at the location where the accident occurred. His report was then forwarded for registration and entry into accident statistics. In some cases, the occurrence of an accident had led to the taking of safety measures.

Investigation procedure

An investigation group consisting of four people was set up at each mill. These were the safety engineer, the safety technician, the trade union safety representative and the production section manager.

The principles of Deviation Investigation (Section 7.5) were applied, albeit in simple form. The first investigations were made with the help of a research specialist in the field.

A total of 12 accidents and three near-accidents were investigated in the course of the experiment, making a total of 15 cases. The investigations themselves each took about 30 minutes. Including meetings and preparing reports on safety measures, the total time devoted to each case came to around one hour.

Results

The quality of investigations was improved. In comparison with previous investigations:

* The amount of information on events preceding an accident was doubled. (From 1,8 to 3.5 deviations per accident.)
* Five times as many safety measures were proposed as a mean value per accident. (From 0.6 to 3 proposals per accident.)

At paper mills, working with paper-rolling machines is particularly hazardous. The frequency of accidents is well above the average for the paper industry as a whole. During the experimental period, priority was given to accidents at these machines, and a total of seven accidents were investigated.

The result was that the number of accidents at paper-rolling machines fell by a third. The number of days of sick leave resulting from such accidents also fell—from an annual average of 420 days to 130. There was also a reduction in the frequency of accidents occurring in the course of other activities at the paper mill, but this cannot be attributed with the same degree of certainty to the accident investigation routine.

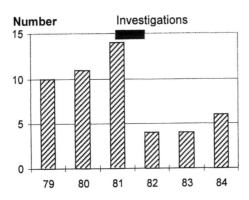

Figure 15.6 Accidents at paper-rolling machines in two paper mills before and after a field experiment (over the years 1979 – 1984).

Economic appraisal

Investigations

The trial activities took up working time of about two man weeks in total, and some extra help to get started. This was considered to be reduced by the educational value of participation. The net cost was estimated at 3000 US dollars.

Investments

Costs were mainly directed at the paper-rolling machines, and half of the time spent on experimental activities was devoted to them. Technical measures cost about 4000 US dollars. The greatest effort concerned an examination of the control system for one of the machines. Deficiencies in the system had caused problems, and had led to both accidents and near-accidents. Solving these problems took up four man weeks, but such an examination would have been carried out sooner or later. Under Alternative II, a cost for this was estimated at 6000 US dollars.

Table 15.5   Costs and benefits of an accident investigation routine (1000 US dollars).

Activity	Alternative I	Alternative II
Investments		
Investigations (net)	– 3	– 3
Technical changes	– 4	– 4
Examination of control system	0	- 6
Total	– 7	– 13
Operations (per year)		
Fewer accidents	+ 6	+ 30
Annual total	+ 6	+ 30

Operations

After the experiment, the number of accidents fell by an average of eight a year, and there were 300 fewer days of sick leave as a result. The company estimated an accident to result in an average cost of around 700 US dollars. Another conceivable way of calculating this cost is to attribute 100 US dollars to each day of sick leave taken (Alternative II). These are the marginal expenses incurred by the company. There are also costs for accidents, which are carried by the social insurance authorities and by the injured persons, and these are not included in the table.

There were no clearly identified benefits for production, although the improved control system for the rolling machine may have been of financial utility.

Total

In this case, earnings were calculated over a period of three years, as major changes were then made to one of the installations. A discount rate of 10% has been assumed, giving a conversion factor of 2.5 (Table 13.12). The current value of the future income attributable to the investment can be estimated at 15 000 or 75 000 US dollars depending on the alternative chosen.

The safety analysis was profitable even in this case, despite the fact that the "only" gain came from fewer accidents and the shorter appraisal period.

Comments

The experiment was successful in terms of its outcome. However, there were a number of problems, and it did not particularly go well according to plan. Examples of problems were:

- Production section managers attended only a few of the meetings.
- Investigations took place a long time after the accidents occurred (40 days on average instead of three days as planned).
- It was sometimes difficult to get safety proposals implemented.

It was intended that the experiment should last for a period of six months; in fact, it ended immediately after that period expired. A decision was made not to reinstate the routines, and this applied even one year later when results had demonstrated a considerable reduction in the number of accidents. The explanation presented to the author was that the company wanted to wait for the reaction of the Paper Industry Work Environmental Council. Since no reaction was forthcoming, activities were not recommenced within the company.

It is rather surprising that the fall in the number of accidents was so great, and that the lower accident frequency was maintained throughout the three years for which evaluation was possible. The results can only be partly explained by the control measures taken. No study of possible explanations has been conducted. Nevertheless, comments made at the end of the experimental period hint at some possible reasons:

> *"Now we have a far better language with which to discuss accidents."*

> *"It was awkward that supervisors got irritated when asked about accidents. Their view was that they had carried out an investigation and everything had been cleared up."*

The first comment refers to people beginning to apply the ideas that lie behind Deviation Analysis, e.g. that there is always a pre-history to the occurrence of an accident.

The second refers to the interest in accidents that had been generated by the experiment. Previously, an accident investigation had been regarded just as a piece of paper that had to be sent off. During the experiment an investigator arrived and posed questions. It may be supposed that this initiated a process through which greater attention was paid to accidents. Attitudes towards them probably changed, above all among job supervisors. This may well be the most important reason why the number of accidents fell so dramatically.

Accident investigation is not a form of safety analysis in a strict sense, since only part of a system and just a few hazards are covered. The "selection" of cases is made at random, i.e. by accidents that have occurred. Deviation theory suggests that "organisational deviations" lie behind most accidents. Such deviations can be rather general by nature, which explains why investigations can provide a type of knowledge applying to a whole installation. A precondition, however, is that a systematic method is employed.

15.7 OVERVIEW ANALYSIS OF A CHEMICALS PLANT

Background

The management of a chemicals plant commissioned a summary risk analysis. The aim was to obtain an overall picture of hazards at the installation.

Analysis procedure

The level of ambition of the analysis was determined by the requirement that the work should be conducted in one week. This meant that some form of "quick" analysis had to be conducted. This was carried out in the form of two meetings of a study team, each lasting around four hours.

The first meeting
It was decided at the first meeting that the entire installation should be examined. To start with, all participants noted down the serious hazards of which they were aware. A list of ten in total was obtained.

A quick form of analysis, based on a combination of the principles of Energy Analysis and Deviation Analysis, was chosen. The object was structured using a block diagram. The types of consequences to which attention should be paid were:

- Fires.
- Explosions.
- Serious occupational accidents.
- Damage to the environment.
- Disturbances to production.

Around 100 different hazards were identified at the first meeting

The second meeting

Hazards were then classified, and a direct risk assessment approach (Table 4.3) was applied. In addition, a category was reserved for recommendation of further investigation or safety analysis. In some cases, it was possible to estimate the financial consequences of an accident.

Twelve potential accidents were picked out, and safety proposals were produced in relation to them.

The most serious plausible accident scenario was a fire in a distillation column for solvents. Such a fire might entail the deaths of five people. An accident of this kind might be triggered off by loss of electrical power, causing overheating and an explosion if corrective measures are not taken immediately within a couple of minutes.

Comments

Around 100 different hazards were identified and about 30 safety proposals made, some of which were general by nature.

A simple analytical method was employed but a large number of hazards were identified and proposals made. Succeeding with such an approach depends much on the study team's knowledge of the plant. The role of the analyst (and the analysis) was to stimulate the imagination of team members, summarise their collective knowledge, and then present it in a systematic manner.

15.8 COMPARISON OF RESULTS FROM THREE METHODS

Background

For this example a number of different methods were applied to the same equipment. The aim of this summary is to illustrate in some detail the different types of results that can be obtained from Deviation Analysis, Energy Analysis, and Safety Function Analysis.

Short summaries of how the methods were applied and a comparison of results are presented below.

The analysed system

The system is the one used in Chapter 10.7 to give an example of Safety Function Analysis. The main features of the system are briefly described again.

The technical part of the production system consists of five similar production tanks, each with a volume of about 3 m^3. These are used to mix various compounds, and no chemical reactions should occur. The site also accommodates a cleaning system using lye and hot water, which is run by a computer-control system. This type of production is common in the pharmaceuticals, food and other similar process industries.

In principle, simple batch production is involved, where different substances are added and mixed following strict procedures. Hygienic demands on the product are high, and cleaning follows specified routines. An essential part of the work is manual, and guided by formal procedures and batch protocols. Twenty people are employed in the workplace, and production is run in shifts.

The workplace is a part of a large factory, with an organisational hierarchy. This means that overall production planning also sets guidelines for health and safety work. The workplace was new, and partly based on a novel design concept. Production had only recently started.

General planning of the analysis

The design of a similar production system was in progress. The aim of the safety analyses was to study the workplace in order to find design improvements for the new production site.

Energy Analysis and Deviation Analysis were chosen as complementary methods. After these had been performed, it was decided to supplement results using Safety Function Analysis.

Evaluation of risks was performed using the direct risk assessment approach (tables 4.3 and 4.4). The assumption was that the assessment concerned the building of a similar but new system. This meant that more improvements were considered than those that would have applied if they applied solely to an existing system.

Performing the safety analyses

The Energy Analysis was performed by the author in collaboration with a safety engineer at the company, who had good knowledge of the technical functions of the system. An evaluation of hazards was performed, and ideas for improvements were proposed. This was done in two meetings, each of around three hours.

The Deviation Analysis was performed in a work group comprising a supervisor and an operator at the installation, and also the safety engineer. The author acted as chairman of the work group. At a first meeting (of about three hours) a list of deviations was produced. During a second meeting the deviations were evaluated, and measures proposed—mainly based on the views of the supervisor and the operator.

The Safety Function Analysis was performed through collaboration with the same persons, plus a production manager with specific system responsibility. For practical reasons, only a restricted part of the system was examined. It was decided to limit analysis to hazards related to the cleaning functions of the system, such as lye, hot water and overpressure. The analysis and the results are described in Section 10.7.

As a first step, a structured list of safety functions was generated. The efficiency and importance of all the items on the list were then assessed. Finally, two safety engineers evaluated whether the system was safe enough or whether improvements were needed.

General results of the safety analyses

Some general information about the analyses is given in Table 15.6. The first row of the table shows the work involved in the safety analyses themselves. There was a need for two or three meetings, each taking between two and three hours. Preparing records of results also took time, about twice that of a meeting. About one day was needed for the Energy Analysis, and just over two days for the Deviation Analysis. The Safety Function Analysis was performed as part of a research project, meaning that extra time was devoted to it. In any normal application, a Safety Function Analysis can be expected to take somewhat longer than a Deviation Analysis.

Table 15.6   Summary of data from three safety analysis.

Description	Method of Analysis			
	Energy	Deviation	Safety Function	All analyses
1. Number of meetings	2	2	3	7
2. Number of identified items	34	56	54	144
3. Items, not acceptable	21	34	37	92
4. Items, not acceptable due to production aspects	8	24	3	35
5. Proposals for actions, total	23	48	47	118
6. Actions, including further investigation	13	21	15	49

One measure of results consists of the number of identified items, which obviously is of a quite different nature. The second row of Table 15.6 shows that 34 energies, 56 deviations and 54 safety functions were found—a total of 144 items. Row 3 shows how many of these were related to an unacceptable level of risk, requiring some kind of improvement.

Types of hazards

One category of hazardous situations was connected with lye and hot water, which could cause serious injuries. These fluids were also capable of giving rise to high pressure under certain conditions. Explosions could not be ruled out, since the tanks were not designed to withstand high pressure. Other hazards were related to falls from height, poor ergonomics, errors in follow-up procedures, and so on.

Thirteen health-related and ergonomic problems were identified, but nothing was found in relation to the environment.

Several of the items were related to production problems (see Row 4 of Table 15.6). Deviation Analysis revealed an especially large number of production-related problems. Of all deviations judged as not acceptable, 70% were related to production, some also in combination with safety.

Analysis of proposals

Results from the application of the three methods can be compared in several ways. The methods identify energies, deviations and safety functions respectively, which means that results are not directly comparable. However, one suitable object for comparative analysis of results lies in the proposed improvements generated by the different methods.

The final two rows of Table 15.6 summarise proposals for action. A total of 118 actions were proposed, most of which were specific proposals for improvement. But as many as 41% of the proposals referred to a need for some kind of further investigation. The main reason given for further investigation was that insufficient knowledge was available on the system, e.g. with regard to computer control. The uncertainty created a requirement for more data to be gathered before final evaluation could be made, which was then to be left for a later occasion.

All three analyses generated proposals for improvements to the system. In order to obtain a more useful summary of results, items were grouped into four main categories:

1. Mechanical.
2. Control system.
3. Management.
4. General or other.

Each proposed measure from any one method was categorised and compared with results from the other methods. If two proposals were judged as having an identical purpose, a special note was made of this. The classifications were then utilised to compile a package of proposals for the company, and also for inter-method comparison (see tables 15.7, 15.8 and 15.9).

Results of the three methods

Table 15.7 provides an overview of the proposals generated by all three rounds of analysis. Clear differences between the methods can be observed. Deviation Analysis generated many proposals for the control system, whereas Safety Function Analysis prompted three times as many suggestions related to management as the two other methods put together.

Table 15.7   Number of proposals generated by each safety analysis method, by category and subcategory of improvements.

Category of improvements	Energy Analysis	Deviation Analysis	Safety Function	Total
Mechanical	**14**	**13**	**5**	**32**
1 Ergonomics, workplace design	10	7	0	17
2 Other	4	6	5	15
Control system	**2**	**19**	**10**	**31**
1 General investigation	0	9	6	15
2 Direct improvement	2	9	4	15
3 Other	0	1	0	1
Management	**0**	**9**	**27**	**36**
1 Instructions for operators	0	5	14	19
2 Routines in the department	0	4	8	12
3 Company level	0	0	5	5
General or other	**7**	**7**	**5**	**19**
1 Handling hazards with lye, etc.	4	4	2	10
2 Other	3	3	3	9
Total	23	48	47	118

It should be remembered that the Safety Function Analysis did not cover the entire system. This might, for example, explain the lack of proposals in the area of ergonomics.

A judgement was made concerning the extent to which the proposed measures might have been formulated at the system-design stage. An overall conclusion is hard to draw, but it was indicated that 70% of the problems would have been possible to identify during design.

Comparing Energy Analysis and Deviation Analysis

Energy Analysis and Deviation Analysis were the methods originally selected for the overall safety analysis. A central issue is how results from the two methods overlap, and also what types of measures they address. The issue of complementary analyses has been discussed in Section 12.4, and the principle is outlined in figures 12.1 and 15.7.

Table 15.8   Numbers of proposed measures from Energy Analysis and Deviation Analysis.

Method/Combination	Category of proposed measures				
	Mecha-nical	Control system	Mana-gement	General	Total
Energy Analysis only	8	0	0	4	12
Deviation Analysis only	8	12	6	4	30
Both methods	3	1	0	1	5
Total	19	13	6	9	47
One method only	16	12	6	8	42

Table 15.8 shows the number of proposals in different categories generated by the two methods. There is a rather small overlap, with only five out of a total of 42 proposals coming from the applications of both methods. It is clear that the two methods tend to support different categories of improvements. In particular, Deviation Analysis proved more effective in identifying control-system and organisational needs.

Overlap between three methods

The set of combinations for all three methods is slightly more complicated. The result is shown in Table 15.9. Eliminating overlaps, the total number of proposals comes to 94, which means that the number of duplicate proposals was 24. Only four proposals were generated by all three methods. These were connected with emergency equipment, overpressure in the tanks, and the blocking of machine movements.

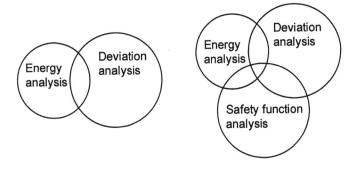

Figure 15.7 Overlap between coverage of methods for safety analysis.

Table 15.9   Number of proposed measures for all three methods by category.

Method/combination	Category of proposed measures				
	Mecha-nical	Control system	Mana-gement	General	Total
Energy Analysis only	8	0	0	4	12
Deviation Analysis only	8	12	6	4	30
Safety Function Analysis only	1	4	24	3	32
Two methods only	6	6	3	1	16
All three methods	1	1	0	2	4
Total, excluding duplicates	24	23	33	14	94
Total, including duplicates	32	31	36	19	118

Some comments

In this case study there were a total of 144 identified items, all of which needed to be evaluated with regard to whether conditions were acceptable or not. Around half-an-hour's meeting time was devoted to evaluations for each analysis. In total, this meant 1.5 hours, giving a mean value of between one and two minutes per item evaluation.

At nearly all evaluations, it was possible to reach consensus. But this was not a compulsory procedure, since it was possible simply to note a dissenting opinion on the analysis record. There was discussion on a few occasions, e.g.

with regard to the dependability of the control system. If sufficient information was not available, some kind of further investigation tended to be proposed.

Three main types of uncertainties became evident during the evaluations:

- What might the consequences be?
- What is the likelihood of something happening? (Usually, this question was posed in relation to the control system and interlocks.)
- How does the system actually work? (Also, a question often related to the control system.)

Rather than making an estimation that would only be a guess of some kind, a further investigation was often recommended to deal with uncertainties. This also operated as a means of telling the reader of an analysis record that information was uncertain. Naturally, such clarification can be especially important if that reader is making decisions related to the design of a new but similar system.

Also, other analyses of the system had been performed at the design stage. The first was conducted by the contractor of the tank, who had attached a "CE label" to indicate compliance with the EU's Machine Directive (EC, 1989). It was not possible to obtain information on the risk evaluations on which approval of the equipment was based. It would have been interesting to study the large differences between the risk assessments of the contractor and those made in this study.

An analysis based on the "What-if" method (see Section 11.7) was performed rather early during the design phase by another team. It focused on dust explosions, but also addressed wider issues. The analysis did not result in any proposals for improvement

Summary

Some lessons may be learned from this case, but it is hard to say how generally valid the conclusions are. Results can be summarised under a few points:

- A large number of improvements (almost 100) were suggested.
- The methods gave clearly different types of results.
- Due to differences in scope, there was only small overlap in results from the three methods. Combining three separate methods proved fruitful.
- Several means for production improvements were identified—especially from the application of Deviation Analysis.

15.9 A QUICK ANALYSIS OF A PRODUCTION LINE

Background

A production line containing several sheet-metal presses was about to be installed. Some parts of the line required detail design. The safety engineer at the company had not previously been involved in the project. He was invited to a meeting with the supplier and project leader to assess possible hazards before the design of these parts was finalised.

Analysis

The safety engineer proposed that a simple safety analysis should be conducted. His suggestion met with a favourable reception. He started with a half-hour presentation of the principles of Energy Analysis and Deviation Analysis. A tour was then made of the site and an immediate "quick" analysis conducted. This took one and a half hours. Drawings of the design proposals were used to supplement the analysis. The group reassembled and went through their notes.

Results

Together they found 30 or so mechanical hazards. Proposals were mainly directed at safety devices, such as machine guards, railings and warning equipment. The costs of the measures were calculated, and an immediate decision on implementation was taken at the meeting.

During the analysis the safety engineer posed a number of questions concerning the computer-based control system, which none of the participants could answer. There appeared to be a number of potential problems on which nobody had a real grip. These problems were later raised with a consultant.

The methodological perspective provided by safety analysis enabled the safety engineer to engage in meaningful dialogue with the consultant, despite the fact that the control system lay outside his own area of expertise.

16
Concluding remarks

Safety analysis is becoming more common, and is starting to be an established way of promoting occupational safety. The area will probably be further developed. Valuable contributions may come from the evaluation of results, and greater emphasis on quality aspects and criteria for risk assessment.

The focus of the book has been on occupational accidents and what can be done about them at company level. But the principles are also applicable to a broader spectrum of systems and undesired events. It would certainly be of advantage if experiences and knowledge of applications could be exchanged across a wider area.

I would like to stress one final time that analyses are best conducted in a team. It is advantageous to apply an integrated approach to safety analysis. This will also mean that economic arguments (which are often powerful) can be employed to back up safety proposals. Safety is not always an expense. It can be good business as well (as said many times before).

Since this guide is designed to cover several methods and different areas of application, its scope is fairly extensive. Using safety analysis can appear demanding at first sight. But it need not be particularly difficult in practice. In many cases, extensive planning is not required, and it is enough to be familiar with just a few methods. When you try safety analysis for the first time, start with a simple approach, giving a preliminary overview of hazards.

The first step is to define the aim of analysis. Why do you need a safety analysis? Do you apply a technical perspective, or include human, organisational and production factors? The next step is to choose one or two methods that will support you in achieving your goal.

It is probably only when you have conducted a safety analysis yourself that you recognise the benefits of this way of working. You detect hazards that would otherwise have remained undiscovered. This can be a stimulating experience!

17
References

Amendola, A., Contini, S., and Ziomas, J., 1992. Uncertainties in chemical risk assessment; Results of a European benchmark exercise. *Journal of Hazardous Materials*, **29**, 347–363.

Andersson, R., 1991. The role of accidentology in occupational injury research. National Institute of Occupational Health, Solna, Sweden.

Annet, J. and Stanton, N.A. (eds), 2000. *Task analysis.* Taylor & Francis, London.

Annet, J., Duncan, K.D., Stammers, R.B. and Gray, M.J., 1971. Task Analysis. Her Majesty's Stationery Office, London.

Association of Swedish Chemical Industries, 1996. SHE-Audit – A guideline for internal auditing of Safety/Health/Environment. AB Industrilitteratur, Stockholm.

Backström, T. and Harms-Ringdahl, L., 1986. Mot säkrare styrsystem – Om personsäkerhet vid automatiserad produktion. Royal Institute of Technology, Stockholm.

Bartlett, F.C., 1932. *Remembering: A study in experimental and social psychology.* Cambridge University Press, Cambridge.

Beatson, M. and Coleman, M., 1997. International Comparisons of the Economic Costs of Work Accidents and Work-Related Ill-Health. In Mossink, J. and Licher, F. (eds): *Proceedings of the European Conference on Costs and Benefits of Occupational Safety and Health 1997,* The Hague.

Bell, J.B. and Swain, A.D., 1983. A procedure for conducting a human reliability analysis for nuclear power plants. U.S. Nuclear Regulatory Commission, Washington.

Bergman, B., 1985. On reliability theory and its applications. *Scandinavian Journal of Statistics*, **12**, 1-41.

Bird, F.B. and Loftus, R.G., 1976. *Loss control management.* Institute Press, Loganville, Georgia.

Both, R.T., Boyle, A.J., Glendon, A.I., Hale, A.R. and Waring, A.E., 1987. Chase II: The Complete Health and Safety Evaluation Manual for Smaller Organisations. Health and Safety Technology and Management, Birmingham.

Brehmer, B., 1987. The psychology of risk. In Singleton, W.T. and Hovden, J. (eds): *Risk and decisions.* Wiley & Sons Inc, New York.

British Safety Council, 1988. Five Star Health and Safety Management Audit System. British Safety Council, London.

Britter, R.B., 1991. The evaluation of technical models used for major-accident hazard installations. University of Cambridge, Cambridge.

Brody, B., Letourneau, Y. and Poirier, A., 1990. An indirect cost theory of work accident prevention. *Journal of Occupational Accidents*, **13**, 225–270.

Brown, D.M. and Ball, P.W., 1980. A simple method for the approximate evaluation of fault trees. *3rd International Symposium on Loss Prevention and Safety Promotion in the process industries.* European Federation of Chemical Engineering.

BSI (British Standards Institution), 1996. Guide to Occupational Health and Safety Management Systems (British standard BS8800: 1996). British Standards Institution, London.

Bullock, M.G., 1976. Change control and analysis. EG&G Idaho Inc, Idaho.

CCPS (Center for Chemical Process Safety), 1985. *Guidelines for Hazard Evaluation Procedures.* American Institute of Chemical Engineers, New York.

CCPS (Center for Chemical Process Safety), 1993. *Guidelines for Safe Automation of Chemical Industries.* American Institute of Chemical Engineers, New York.

CEC, 1989. Introduction of measures to encourage improvements in the safety and health of workers at work (Directive 89/391/EC). Council of the European Communities, Brussels.*

CEC, 1989/98. Laws of the Member States relating to machinery (Directive 89/392/EC and 98/37/EC). Council of the European Communities, Brussels.*

CEC, 1993. The EC Eco Management and Audit Scheme (EMAS). Council of the European Communities, 1836/93, Brussels.*

CEC, 1996. Control of major-accident hazards involving dangerous substances (Council Directive 96/82/EC). Council of the European Union, Brussels.*

CEN, 1996. Safety of machinery – Principles for risk assessment (EN 1050:1996). European Committee for Standardization, Brussels.

Chaplin, R. and Hale, A., 1998. An evaluation of the use of the international safety rating system (ISRS) as intervention to improve the organisation of safety. In Hale, A. and Baram, M. (eds): *Safety management - The challenge of change.* Elsevier Science, Oxford.

CISHC (Chemical Industry and Safety Council), 1977. *A Guide to Hazard and Operability Studies.* Chemical Industries Association, London.

Clemens, P.E., 1982. A compendium of hazard identification and evaluation techniques for system safety application. *Hazard Prevention*, **18**, 11-18.

Collins English Dictionary, 1986. William Collins Sons and Co. Ltd, London.

---

*Full report available on the Internet; address at http://www.irisk.se/ref.htm.

Committee on Trauma Research, 1985. *Injury in America*. National Academy Press, Washington.

Contini, S., Amendola, A. and Ziomas, I., 1991. Benchmark Exercise on Major Hazard Analysis. Commission of the European Communities, Joint Research Centre, Ispra, Italy.

Cox, S. and Tait, R., 1998. *Safety, Reliability and Risk Management* (2nd edn). Butterworth-Heinemann, Oxford.

Dansk standard. 1993. Risk analysis: requirements and terminology (in Danish). Dansk standard, Copenhagen.

DNV (Det Norske Veritas), 1990. Introduction to International Loss Control Institute and the International Safety Rating System. DNV Industritjenster A/S, Oslo, Norway.

Dorman, P., 2000. The Economics of Safety, Health, and Well-Being at Work: An Overview. International Labour Office, Geneva*.

Eisner, J. and Leger, J.P., 1988. The international safety rating system in South African mining. *Journal of Occupational Accidents*, **10**, 141–160.

Embrey, D., 1994. *Guidelines for Preventing Human Error in Process Safety*. American Institute of Chemical Engineers, New York.

Engineering Council, 1993. Guidelines on Risk Issues. The Engineering Council, London.

EPA, 1990. The Clean Air Act Amendments of 1990. The United States Environmental Protection Agency, Washington.*

Farmer, E. and Chambers, E., 1926. A psychological study of individual differences in accident rates. Industrial Health Research Board, Report No 38, London

Ferry, T.S., 1988. *Modern Accident Investigation and Analysis* (2nd edn). John Wiley & Sons Inc, New York.

Fischoff, B., Lichtenstein, S., Lovic, P., Derby, L. and Keeney, R., 1981. *Acceptable risk*. Cambridge University Press, Cambridge.

Freud, S., 1914. *Psychopathology of everyday life*. Ernest Benn, London.

Gertman, D.I. and Blackman, H.S., 1994. *Human Reliability and Safety Analysis Data Handbook*. John Wiley & Sons, New York.

Gibson, J.J., 1961. Contribution of experimental psychology to the formulation of the problem of safety: a brief for basic research. In *Behaviour Approaches to Accident Research*. Association for the Aid of Crippled Children, New York, 77–89.

Gordon, J.E., 1949. The epidemiology of accidents. *American Journal of Public Health*, **9**, 504–515.

Grimaldi, J., 1947. Paper at the ASME standing committee on Safety. Atlantic City, N.J.

Guastello, S.J., 1991. Some further evaluations of the International Safety Rating System. *Safety Science*, **14**, 253–259.

Haddon, W. Jr., 1963. A note concerning accident theory and research with special reference to motor vehicle accidents. *Annals of New York Academy of Science*, **107**, 635–646.

Haddon, W. Jr., 1980. The basic strategies for reducing damage from hazards of all kinds. *Hazard Prevention*, **16**, 8–12.

Hale, A.R., 1990. Safety Rules O.K.? Possibilities and limitations in behavioural safety strategies. *Journal of Occupational Accidents*, **12**, 3–20.

Hale, A. and Glendon, I., 1987. *Individual behaviour in the control of danger*. Elsevier, Amsterdam.

Hale, A.R., Heming, B.H.J., Carthey, J. and Kirwan, B., 1997. Modelling of safety management systems. *Safety Science*, **26**, 121–140.

Hammer, W., 1972. *Handbook of system and product safety*. Prentice Hall Inc., New Jersey.

Hammer, W., 1980. *Product safety management and engineering*. Prentice Hall, Inc., New Jersey.

Harms-Ringdahl, L., 1982. Riskanalys vid projektering – Försöksverksamhet vid ett pappersbruk. Royal Institute of Technology, Stockholm.

Harms-Ringdahl, L., 1983. Vis av skadan – Försök att systematiskt utreda och förebygga olycksfall vid två pappersbruk. Royal Institute of Technology, Stockholm.

Harms-Ringdahl, L., 1986. Experiences from safety analysis of automatic equipment. *Journal of Occupational Accidents*, **8**, 139–146.

Harms-Ringdahl, L., 1987a. Safety analysis in design – evaluation of a case study. *Accident Analysis and Prevention*, **19**, 305–317.

Harms-Ringdahl, L., 1987b. *Säkerhetsanalys i skyddsarbetet – En handledning*. Folksam, Stockholm. (Original version of this book.)

Harms-Ringdahl, L., 1990. On economic evaluation of systematic safety work at companies. *Journal of Occupational Accidents*, **12**, 89–98.

Harms-Ringdahl, L., 1991. Att förebygga olycksfall – Säkerhetsanalys i företagshälsovården. Swedish Work Environment Fund, Stockholm.

Harms-Ringdahl L., 1999. On the modelling and characterisation of safety functions. In Schueller, G.I. and Kafka, P. (eds). *Safety and Reliability ESREL'99*. Balkema, Rotterdam, 1459–1462.

Harms-Ringdahl, L., 2000. Assessment of safety functions at an industrial workplace – a case study. In Cottam, M.P., Harvey, D.W., Pape, R.P., and Tait, J. (eds): *Foresight and Precaution, ESREL2000*. Balkema, Edinburgh, 1373–1378.

Health and Safety Executive, 1989. *Human factors in industrial safety*. Health and Safety Executive, London.

Health and Safety Executive, 1991. *Successful Health and Safety Management*. Health & Safety Executive, London.

Heinrich, H.W., 1931. *Industrial Accident Prevention*. McGraw-Hill, New York.

Heinrich, H.W., Petersen, D. and Roos, N., 1980. *Industrial Accident Prevention* (5th edn). McGraw-Hill, New York.

Hendrick, K. and Benner, L., 1987. *Investigating accidents with STEP*. Marcel Dekker Inc, New York.

Henley, J.H. and Kumamoto, H., 1981. *Reliability engineering and risk assessment*. Prentice-Hall Inc., New Jersey.

Hollnagel, E., 1993. *Human Reliability Analysis Context and Control*. Academic Press, London.

Hollnagel, E., 1998. *Cognitive Reliability and Error Analysis Method –CREAM*. Elsevier Science, Oxford, England.

Hollnagel, E., 1999. Accident Analysis and Barrier Functions. Institute for Energy Technology, Kjeller, Norway.

Hollnagel, E., 2000. On understanding risks: Is human reliability a red herring? In Svedung, I. (ed). ESReDa-seminar: Risk Management and Human Reliability in Social Context, Karlstad, Sweden.

Hurst, N.W., Young, S., Donald, I., Gibson, H. and Muyselaar, A., 1996. Measures of safety management performance and attitudes to safety at major hazard sites. *Journal of Loss Prevention in the Process Industries*, 9 (2), 161-172.

IAEA, 1994. ASCOT Guidelines: Guidelines for self-assessment of safety culture and for conducting review. International Atom Energy Agency, Vienna, Austria.

IEC (International Electrotechnical Commission), 1985. Analysis techniques for system reliability – Procedure for failure mode and effects analysis (IEC 812). IEC, Geneva.

IEC (International Electrotechnical Commission), 1990. Analysis techniques for system reliability – Procedure for fault tree analysis (IEC 1025). IEC, Geneva.

IEC (International Electrotechnical Commission), 1995. Dependability management – Risk analysis of technological systems (IEC 300-3-9). IEC, Geneva.

IEC (International Electrotechnical Commission), 1998. Functional safety: safety related systems (IEC 1508). IEC, Geneva.

ILO (International Labour Organisation), 1988. *Major Hazard Control. A practical manual*. International Labour Office, Geneva.

INSAG (International Nuclear Safety Advisory Group), 1988. Basic safety principles for Nuclear Power Plants. International Atomic Energy Agency, Vienna.

INSAG (International Nuclear Safety Advisory Group), 1996. Defence in depth in nuclear safety. International Atomic Energy Agency, Vienna.

ISO, 1987. Quality management and quality assurance standards – Guidelines for selection and use (ISO 9000). International Standardisation Organization, Geneva.

ISO, 1990. Guidelines for auditing quality systems – Part 1: Auditing (ISO 10 011). International Organization for Standardization, Geneva.

ISO, 1996. Environmental Management System – Specification with the guidance for use (ISO 14001). International Organization for Standardization, Geneva.

Jenkins, A.M., Brearley, S.A. and Stephens, P., 1991. Management at risk. AEA Technology, Cheshire, England.

Johnson, W.G., 1980. *MORT Safety Assurance Systems*. National Safety Council, Chicago.

Karolinska Institutet, Department of Social Medicine. The World Health Organization Manifesto for safe communities. 1st World Conference on Accident and Injury Prevention. Stockholm, 1989.

Kecklund, L., Edland, A., Wedin, P. and Svenson, O., 1995. Comparison of safety barrier functions in the refueling process in a nuclear power plant before and after a technical and organizational change. In Norros, L. (ed): *Fifth European Conference on Cognitive Science Approaches to Process Control*, VTT, Espoo, Finland.

Kennedy, R. and Kirwan, B., 1998. Development of a Hazard and Operability-based method for identifying safety management vulnerabilities in high risk systems. *Safety Science*, **30**, 249–274.

Kepner, C.H. and Tregoe, B., 1965. *The rational manager*. McGraw-Hill, New York.

Kirwan, B., 1994. *A guide to practical human reliability assessment*. Taylor & Francis Ltd, London.

Kirwan, B. and Ainsworth, L.K. (eds), 1993. *A Guide to Task Analysis*. Taylor & Francis, Washington, D.C.

Kjellén, U., 1984. The deviation concept in occupational accident control – Definition and classification. *Accident Analysis and Prevention*, **16**, 289–306.

Kjellén, U., 2000. *Prevention of accidents through experience feedback*. Taylor & Francis, London.

Kletz, T.A., 1983. HAZOP and HAZAN. Notes on the identification and assessment of hazards. The Institution of Chemical Engineers, Rugby, England.

Kliesch, G., 1988. ILO report by the chief of the occupational safety and health branch. *Proceedings XIth World Congress of the Prevention of Occupational Accidents and Diseases*. Stockholm.

Know, N.W. and Eicher, R.W., 1976. MORT user's manual. For use with the Management Oversight and Risk Tree analytical diagram. EG&G Idaho Inc, Idaho.

Krug, E. (ed.), 1999. Injury: A Leading Cause of the Global Burden of Disease. World Health Organization, Geneva. *

Kumamoto, H. and Henley, E.J., 1996. *Probabilistic Risk Assessment and Management for Engineers and Scientists* (2nd edn). IEEE Press, New York.

---

* *Full report available on the Internet; address at http://www.irisk.se/ref.htm.*

Lees, F. P. 1996. *Loss prevention in the process industries* (2nd edn). Butterworth-Heinemann, Oxford.

Leigh, J.P., Markowitz, S., Fahs, M., Shin, C. and Landrigan, P., 1996. Costs of Occupational Injuries and Illnesses (NIOSH Report U60/CCU902886). National Institute for Occupational Safety and Health, USA.

McAfee, R.B. and Winn, R., 1989. The use of incentives/feedback to enhance work place safety: A critique of the literature. *Journal of Safety Research*, 2, 7–18.

McElroy, F. (ed.), 1974. *Accident Prevention Manual for Industrial Operations* (7th edn). National Safety Council, USA.

Nielsen, D.S., 1971. The cause consequence diagram method as a basis for quantitative accident analysis (Report Risö-M-1374). Risö National Laboratory, Denmark.

Nielsen, D.S., 1974. Use of cause-consequence charts in practical systems analysis (Report Risö-M-1743). Risö National Laboratory, Denmark.

Norsk standard, 1991. Requirements for risk analysis. Norges Standardiserings-forbund, Oslo.

O'Connor, P.D.T., 1991. *Practical reliability engineering* (3rd edn). Wiley & Sons Inc, New York.

Perrow, C. 1984. *Normal accidents – Living with high-risk technologies.* Basic Books (2nd edn, 1999, Princeton University Press, Princeton, USA).

Petersen, D., 1982. *Human error reduction and safety management.* Garland STPM Press, New York.

Pitbaldo, R.M., Williams, J.C. and Slater, D.H., 1990. Quantitative assessment of process safety programs. *Plant Operations Progress*, 9 (3), 169–175.

Rasmussen, J. and Jensen, A., 1974. Mental procedures in real-life tasks: A case study of electronic troubleshooting. *Ergonomics*, 17, 293–307.

Rasmussen, J., 1980. What can be learned from human error reports. In Duncan, K.D., Gruneberg, M. and Wallis, D. (eds): *Changes in Working Life.* Wiley & Sons Inc, New York.

Reason, J., 1990. *Human error.* Cambridge University Press, New York.

Reason, J., 1997. *Managing the risks of organizational accidents.* Ashgate Publishing, Aldershot.

Rouhiainen, V., 1990. The quality assessment of safety analysis. Technical Research Center of Finland, Espoo, Finland.

Rouhiainen, V., 1992. QUASA: A method for assessing the quality of safety analysis. *Safety Science*, 15, 155–172.

Rouhiainen, V., 1993. Modelling of accident sequences. In Suokas, J. and Rouhiainen,V. (eds): *Quality Management of Safety and Risk Analysis.* Elsevier, Amsterdam.

Ruuhilehto, K., 1993. The management oversight and risk tree (MORT). In Suokas, J. and Rouhiainen,V. (eds): *Quality Management of Safety and Risk Analysis.* Elsevier, Amsterdam.

Saari, J., 1990. On strategies and methods in company safety work: From informational to motivational strategies. *Journal of Occupational Accidents*, **12**, 107–117.

Schaff, v d T.W., Lucas, D.A. and Hale A. R., 1991. *Near Miss Reporting as a Safety Tool*. Butterworth-Heinemann Publishing, Oxford.

Schofield, S., 1998. Offshore QRA and the ALARP principle. *Reliability Engineering and System Safety*, **61**, 31–37.

Schurman, D.L. and Fleger, A.F., 1994. Human factors in HAZOPS: guide words and parameters. *Professional Safety*, **39**, 32–34.

SCRATCH, Scandinavian Risk Analysis Technology Cooperation, 1984. *Sikkerhetsanalyse som beslutningsunderlag*. Yrkeslitteratur, Oslo.

Simon, H.A., 1957. *Models of man*. Wiley & Sons Inc, New York.

Sundström-Frisk, C., 1984. Behavioural control through piece-rate wages. *Journal of Occupational Accidents*, **6**, 49-59.

Suokas, J., 1985. On the reliability and validity of safety analysis. Technical research center of Finland, Espoo, Finland.

Suokas, J. and Kakko, R., 1989. On the problems and future of safety and risk analysis. *Journal of Hazardous Materials*, **21**, 105-124.

Suokas, J. and Rouhianen, V., 1984. Work safety analysis. Method description and user's guide. Technical Research Center of Finland, Espoo, Finland.

Suokas, J. and Rouhiainen, V. (eds), 1993. *Quality Management of Safety and Risk Analysis*. Elsevier, Amsterdam.

Svenson, O., 1991. The accident evolution and barrier function (AEB) model applied to incident analysis in the processing industries. *Risk Analysis*, **11**, (3), 499–507.

Svenson, O., 2000. Accident Evolution and Barrier Function (AEB) Method – Manual for Accident Analysis. Swedish Nuclear Power Inspectorate, Stockholm.*

Swain, A.D. and Guttman, H.E., 1983. *Handbook of human reliability analysis with emphasis on nuclear power plant applications*. U.S. Nuclear Regulatory Commission, Washington DC.

Söderqvist, A., Rundmo, T. and Alltonen, M., 1990. Costs of occupational accidents in the Nordic furniture industry (Sweden, Norway, Finland). *Journal of Occupational Accidents*, **12**, 79–88.

Takala, J. 1998. Global estimates of fatal occupational accidents. *Sixteenth International Conference of Labour Statisticians*, International Labour Office, Geneva.*

Taylor, A., 1981. Comparison of actual and predicted reliabilities on a chemical plant. The Institution of Chemical Engineers, IChemE Symposium No 66.

---

* *Full report available on the Internet; address at http://www.irisk.se/ref.htm.*

Taylor, J.R., 1974. Sequential effects in failure mode analysis (Report Risö-M-1740). Risö National Laboratory, Denmark.

Taylor, J.R., 1979. A background to risk analysis. Electronics Department, Risö National Laboratory, Denmark.

Taylor, J.R., 1982. Evaluation of costs, completeness and benefits for risk analysis procedures. Int. symp. on Risk and Safety Analysis, Bonn.

Taylor, J.R., 1994. *Risk analysis for process plan, pipelines and transport.* E & FN Spon, London.

Taylor, J.R., Becher, P., Pedersen, K.E., Kampmann, J., Schepper, L., Kragh, E. and Selig, R., 1989. Quantitative and qualitative Criteria for the Risk Analysis. Danish Environmental Agency, Copenhagen, Denmark.

Vlek, C. and Stallen, J.P., 1981. Judging risks and benefits in the small and in the large. *Organizational Behaviour and Human Performance,* **28,** 235–271.

Wagenaar, W.A., Groeneweg, J., Hudson, P.T.W. and Reason, J. T., 1994. Promoting safety in the oil industry. *Ergonomics,* **37,** 1999–2013.

Wahlström B., 1994. Models, modelling and modellers: an application to risk analysis. *European Journal of Operational Research,* **75,** 447–487.

Wilde, G., 1982. The theory of risk homeostasis: implications for safety and health. *Risk Analysis,* **2,** 209–225.

Wilpert, B. and Fahlbruch, B., 1998. Safety related interventions in Inter-Organisational fields. In Hale, A. and Baram, M. (eds): *Safety management – the challenge of change.* Pergamon, Oxford.

18
Index